Unfolding Spatial Movements in the Second-Hand Book Market in Kolkata

This insightful book unfolds the boipara, exploring the acts of thinking and writing about space and place in the context of recent key conversations at the intersections of cultural geographies, mobilities, materialities and heritage studies.

This book reconsiders how we can think about space, place and spatialisation using the book market as a case study. Focusing on everyday lived and imagined experiences within the space, it provides insights into the intricacies, complexities and mobilities involved in the many ways in which temporal, material, structural and sensorial experiences of spaces are inter-implicated. As expression and method, this work aims to be a writing of space (rather than a writing about space) produced through the interleafing of the author's lived spatial experience of the boipara with the stories, experiences and memories of other regulars who have used and continue to use it, along with the non-human materialities and mobilities that characterise it.

This book is essential reading for a wide international audience, particularly those interested in the evolving discussions on mobility, or writing about space and place, materiality, assemblage theory and heritage spaces in the South Asian context.

Diti Bhattacharya is a Postdoctoral Research Fellow with the Griffith Centre for Social and Cultural Research at Griffith University. She is currently working on an Arc Discovery Project titled 'Engaging Outsiders in Sport: Transforming Major Sport Event Legacy Planning through a Co-Creation Approach'. Her area of research expertise includes human and cultural geography, South Asian diaspora, sporting geographies and leisure and tourism geographies. Her doctoral thesis examined spatial movements and material attachments in the second-hand book market of College Street, Calcutta. She combines her research practice working as a research assistant, sessional lecturer and tutor, and as a freelance writer and photographer for various publications. She also enjoys expressing her experiences as migrant woman of colour in the form of short stories, blogs and other creative practice.

Critical Studies in Heritage, Emotion and Affect

In Memory of Professor Steve Watson (1958–2016)

Series Editors: Divya P. Tolia-Kelly (*University of Sussex*) and Emma Waterton (*Western Sydney University*)

This book series, edited by Divya P. Tolia-Kelly and Emma Waterton, is dedicated to Professor Steve Watson. Steve was a pioneer in heritage studies and was inspirational in both our personal academic trajectories. We, as three editors of the series, started this journey together, but alas we lost his magnificent scholarship and valued counsel too soon.

The series brings together a variety of new approaches to heritage as a significant affective cultural experience. Collectively, the volumes in the series provide orientation and a voice for scholars who are making distinctive progress in a field that draws from a range of disciplines, including geography, history, cultural studies, archaeology, heritage studies, public history, tourism studies, sociology and anthropology – as evidenced in the disciplinary origins of contributors to current heritage debates. The series publishes a mix of speculative and research-informed monographs and edited collections that will shape the agenda for heritage research and debate. The series engages with the concept and practice of Heritage as co-constituted through emotion and affect. The series privileges the cultural politics of emotion and affect as key categories of heritage experience. These are the registers through which the authors in the series engage with theory, methods and innovations in scholarship in the sphere of heritage studies.

People-Centred Methodologies for Heritage Conservation
Exploring Emotional Attachments to Historic Urban Places
Edited by Rebecca Madgin and James Lesh

Heritage is Movement
Heritage Management and Research in a Diverse and Plural World
Tod Jones

Unfolding Spatial Movements in the Second-Hand Book Market in Kolkata
Notes on the Margins in the *Boipara*
Diti Bhattacharya

For more information about this series, please visit: www.routledge.com/Critical-Studies-in-Heritage-Emotion-and-Affect/book-series/CSHEA

Unfolding Spatial Movements in the Second-Hand Book Market in Kolkata

Notes on the Margins in the *Boipara*

Diti Bhattacharya

LONDON AND NEW YORK

First published 2024
by Routledge
4 Park Square, Milton Park, Abingdon, Oxon OX14 4RN

and by Routledge
605 Third Avenue, New York, NY 10158

Routledge is an imprint of the Taylor & Francis Group, an informa business

British Library Cataloguing-in-Publication Data
A catalogue record for this book is available from the British Library

ISBN: 978-1-032-27482-9 (hbk)
ISBN: 978-1-032-27483-6 (pbk)
ISBN: 978-1-003-29302-6 (ebk)

DOI: 10.4324/9781003293026

Typeset in Times New Roman
by SPi Technologies India Pvt Ltd (Straive)

Contents

Figures

Acknowledgement

This book is a culmination of my doctoral research thesis. I would like to express my deep gratitude to my supervisory team, Associate Professor Patricia Wise and Professor Sarah Baker. I consider myself to be extremely lucky to be supervised by you. Throughout this project, I have not only undergone growth and understanding academically but also as a person. Thank you both for inspiring me, keeping faith in me and most importantly caring for me at all times. Through your deep experience in academia, you have mentored and guided me immensely about research, thinking and academic writing. Today, I yearn to learn more and research further, largely because I have had the privilege of working under your guidance. More importantly, you have imbued in me the value of being a patient, persistent, confident and caring. Several times over the last four years, you have believed in my capabilities and ideas when I was close to giving up and today it is because of your faith in me that I have a thesis. Thank you for believing in me and supporting me, unconditionally. There is always that one teacher in your life who plants a seed of thought that never leaves you. Dr Peter Wise was that for me. Thanks, Peter, for your continued theoretical guidance in fulfilling this idea.

I am thankful to the Australian Academy of Humanities Publication Subsidy Scheme 2022 for their generous support in having this manuscript published.

I would like to thank Dr. Robert Mason for being a constant support and source of encouragement, both as a Higher Degree Research Convener and as a friend. Your warmth, mentorship and friendship are very special and dear to me. My sincere thanks to Dr. Adele Pavlidis who is my supervisor and mentor, and her continued support has given me a flicker of hope in perusing and intellectual career in very difficult times. A special thanks to Dr Ben Anderson and Dr Ben Highmore for their generous feedback and examination of my thesis and their emphatic encouragement in publishing it into a book. My heartfelt thanks to the editor of this book Faye Leerink, the series editors Prof. Emma Waterton and Divya Tolia Kelly for accepting my ideas and being patient with me during the publication process of this book.

I acknowledge the continued support and opportunities I have received from the Griffith Graduate Research School, the School of Humanities, Languages and Social Science and the Griffith Centre for Social and

Cultural Research. It is because of their continued support through funding and otherwise that I was able to present my work at a number of significant conferences. I would also like to thank the academic and administrative staff of the School, Centre, the Library and The International Student Support Unit. I am indebted to their continuous guidance, help and support through this journey. A special thanks to international student advisor, Mark Taylor for his support in making my journey as an international student smooth and easy.

To my father, Gautam Bhattacharya, my aunty, Shampa Bhattacharya, my cousin Nupur Bhattacharya and everyone else in my household in Calcutta, nothing I do is possible without you being a part of my life and this thesis is no exception. I am thankful to my in-laws and friends in India for their continued support as well.

To my partner and soulmate, Subhankar – your love, warmth, understanding and support from afar and near is what kept me going. Your love is what keeps me going.

Ma, this thesis is dedicated to you. I know you are watching me and guiding me from somewhere special.

Finally, to Dadu thank you for a wonderful childhood full of stories, memories, laughs, walks and the *boipara*. This thesis is for you.

1 Introduction

Introducing the *boipara*

The second-hand book market of College Street, Calcutta, known as the *boipara*, holds a significant place in the city's cultural and political life in its history, present and imaginary. This book unfolds the space as I have experienced it, attending to movements between the material, the sensory and the human, and between experience, memory and imagination. In undertaking this unfolding, I reconsider how we can think about affective attachments through material entanglements. This book is intended to initiate a conversation about the value of thinking of intangible experiences of spaces of heritage through these sensorial registers. This work is centrally focused on experiencing the *boipara* as a cultural practice, rather than revisiting a cultural idea. While space for me is the animated every day, based on practice and experiment, spatialisation is processual and consequential. By focusing on everyday lived and imagined experiences within the *boipara*, I aim to provide insights into the intricacies, complexities and mobilities involved in the many ways in which temporal, material, structural and sensorial experiences of spaces are inter-implicated.

The *boipara*, or neighbourhood of books, consists of hundreds of makeshift second-hand bookstalls, some of which are removed nightly and replaced each morning. It is a phenomenon that fascinates locals, visitors and people who have only seen images or heard descriptions, but for me the *boipara* is a precinct with which I have a long and intimate familiarity. I was born and lived most of my life in Hatibagan, a neighbourhood that is in close proximity to the College Street area. My family has a very long history of intellectual, political, social and cultural engagement with the precinct. There is therefore no way in which I could for a moment write about the *boipara* as anything other than taking place inside-and-outside at once, in place, in time and in perspective.

In other words, I have used my own intimate and personal experiences of growing up and spending time in and around the College Street *boipara* as my primary source of inspiration and motivation. However, these experiences have in turn become entryways to the *boipara* more broadly and deeply, enabling me to weave my own stories into the experiential knowledges of other users of the

DOI: 10.4324/9781003293026-1

Figure 1.1 The second-hand book market, in College Street, Calcutta, or the *boipara*

space, with which I have been able to engage precisely because of our fragments of shared experience. The stories that emerge from the *boipara* therefore range across multiple subjectivities, temporalities and contexts. Further, throughout the project, I have continually negotiated between being a regular customer of the *boipara* and a researcher, producing a realm of in-betweenness. Like the *boipara*, my work is – in Deleuzian terms – an assemblage: fluid, unsettled, continually open to connections within and beyond itself.

Although the reliance on a combination of stories and experiences about the *boipara* could imply the need for chronology, the book does not intend to trace any timeline. This is primarily because the lived experiences of this space have never made sense to me in any chronological order. Rather, I am trying to write about a space that functions in and through multiple layers of time simultaneously, which can only be dealt with through interdisciplinary and transdisciplinary approaches. The *boipara* is historical because it brings the past, through memories and lived experiences, into dialogue with the present (and the future). It is cultural because it not only deals in books, education and conversation, but enacts mundane spatial practices that impart a distinct character to the city of Calcutta as a whole. It is social because it brings together people of different backgrounds and makes them its own. It is geographical, but also philosophical, because the precinct's spatial movements, through its lanes, by-lanes and footpaths, continually throw its self-narrative into a churn that *is* the every day. Most importantly, its value as a space of heritage, wellbeing and sustainability is grounded in the logic of a circulation – between memory, material mobility and affect. Thus, although we use the topography of the *boipara* every day, we live through its material, affective, sensorial and intellectual impacts on us. For me, the *boipara* is also creative because in its unfolding through stories and experiences, both lived and imagined, the space continually creates opportunities to toy with what might happen next, or what could have been. Through its repeated everyday routines taking place within proliferating multiplicities, the *boipara* functions through subtle differences within those repetitions. The relatively unsettled approach I have taken to providing interwoven narratives reflects an ongoing interaction of spatial experience that is itself unsettled and unsettling, but will doubtless continue as long as Calcutta and the *boipara* continue to be resistant to the replacement of idiosyncratic urban arrangements with shopping malls and office buildings.

The *boipara* and me

A significant part of my growing up took place amid old and new books, magazines and newspapers, and myriad groups of collected and curated reading material. I remember Ma, Baba, Dadu and Dida[1] and the other members of our extended family being emotionally and culturally attached to books, newspapers and magazines. It is one thing to be an avid reader, which they were, but it is quite another to be attached to the reading material physically, beyond its content. I have never seen them throw away old newspapers. At the same time, I have never seen them be fascinated only by new books. Old books held equal and sometimes greater value for them. In fact, I have come to realise that they didn't attach comparative value to the oldness and newness of the books, in particular. Their relationship with books was obviously about the content. They were each, in their own ways emotionally hungry for printed words, yearning to grasp all of them. The medium was not the measure: it was the content and expression that mattered.

Each room in my home in Calcutta had specific spaces or corners where these collections of newspapers, little magazines[2] and books lived. Each family member had certain designated spaces in the house that they preferred to use for reading. While Dadu liked to read his books in his study, Dida liked to sit on the porch on sunny afternoons, drying her hair and voraciously finishing her novels. For Baba, books were his everyday commute-to-work companion, while Ma liked to read at bedtime. However, there was one thing they had in common: they were all advocates of reading second-hand books and sourcing them from the *boipara*, and they had all been regulars of the *boipara* at some time in their lives. Thus, it was never only about the books (*boi* in Bengali) but also about the neighbourhood of books (*boipara*).

Like every other young person in our extended family, at a very early age I was engaged in the customary conversation with Ma and Dadu about the importance of developing a habit of reading books. At that age, I found this to be rather an imposition as I could not understand why anyone would want to read outside of anything prescribed by one's school. Clearly my relationship with books was then quite distant from that of the older members of my family. What came to fascinate me, though, was that their relationship was not only with the content and expression of any given book, but with the materiality of the book itself. It was as if books, magazines and other reading materials lived in our home along with us. I use the term *lived* (as opposed to them being stored or present) because every item of reading material in my house had a life and its life was valued. It is one thing to possess material objects such as books, newspapers and so on for utilitarian or productive reasons; rather, through my growing years I witnessed the extent to which books – old and new – and the other reading materials I have described had a special place in our everyday lives. They were not only used, but taken care of, respected and loved *because they shared knowledge with us*. This is why I felt as if the books did not simply coexist in our house, but that they *lived with us*.

Some of my earliest memories are of spending hours on Sundays and school holidays in my grandparents' drawing room, which doubled as my Dadu's study. There was a special corner designated only for old newspapers and news magazines, organised in order of date of publication. Near it was another area piled high with little magazines and periodicals. Dadu also had a cupboard that ran the full length of one wall, filled primarily with second-hand books. Indeed, I noticed over time that Dadu had a particular penchant for collecting and reading second-hand books, old magazines and newspapers, and for this he undertook frequent and regular visits to the *boipara*. Dadu was a big advocate of reading *everything*. From newspapers to news magazines and from little magazines, periodicals and journals to books, he did not believe in the concept of good content and bad content. He asked, 'Until you read it, how would you know what is useful and what is useless?' He also remarked, 'What may be valueless to you could be valuable information to someone else.'

His eclectic and voracious reading habit meant he had a bit of a reputation as an encyclopaedic elder of the neighbourhood. As a consequence, Dadu's drawing room also served as a regular space for *adda*[3] (evening conversations) with many of our friends and family. A big part of this was due to Dadu's wealth of collected reading materials *and* his in-depth knowledge of the bookstores and distribution of books in the *boipara*. Everyone assumed that if anyone might have an old edition of a rare book or a copy of a months-old newspaper, it would be him. Further, if he did not have it, he would probably know which second-hand bookstall at the *boipara* would likely still have copies. It is evident that his relationship with the second-hand books and other reading materials was not limited simply to collecting them. He had a continuous, mobile, dynamic relationship with the spatiality of the *boipara*, from whence the books and so on came and to which they often returned.

There was a big, black, wooden table set right in the middle of his study. In addition to Dadu, I became a regular reader at this table. Having left behind my initial confusion as to why one would read when one didn't have to do it for school, I was *becoming a reader*. Often perched on top of the table instead of sitting on the chairs around it, I nurtured the seeds of my own reading habit. Of course, apart from reading, I had innumerable opinions and questions around the ways in which Dadu had the room organised. Dadu answered my persistent questions for as long as his patience permitted.

One of my regular weekend activities included reading a paragraph of the editorial sections of old newspapers or the prefaces in old books and picking up different parts of speech. As someone who grew up in tumultuous political times and without the chance to complete school, Dadu believed that one could learn the finer nuances of a language (in my case, it was both English and Bengali) by reading old books and newspaper editorials. I often wondered what it was with old books and not new ones that made them so important for him. When I eventually asked, he explained that old second-hand books give you permission to scribble on them, to create notes in margins – that is, to become part of conversations that others had begun.

As far as I can recall, from my earliest observations I was very inquisitive regarding how and why Dadu had specific, designated spaces in his study/drawing room for the second-hand books, new books, old newspapers, different kinds of magazines and so on. Sitting on the table, I gradually noticed how over the years he had in fact created, established and nurtured a unique relationship between his old and new books, the intimate yet specific spaces where they lived in the room and the spatiality of the *boipara* itself. Often, when his friends visited, they indulged in passionate and excited conversations around politics, everyday life, news, media, poetry and, above all, books. Interestingly, the conversations around books were often not about the content of the books, but about their sources – that is, where they materially came from. References to the *boipara* were frequent. Dadu was often commended (sometimes genuinely and sometimes ironically) by friends and family alike about his love, care and obsession with second-hand books from the *boipara*. He was frequently described by those who knew him best with the Bengali adjective *shoukhin*.[4] The word *shoukhin* has no straightforward equivalent in English. A *shoukhin* person seeks to discover the difference between the essence of orange blossom, lavender and sandalwood as if their life depended on it. It is not, however, a question of investing hours in the finer things of life or of being a pedant. It is a combination of intricate, intimate care and attachment to the ordinary things. His emotional investment lay not only in the fact that he had read all the second-hand books that he had possessed but also in the space from which he had collected them, the *boipara*.

Spending a considerable time in his study, I was gradually able to fathom the importance of the space he had designated to the second-hand materials in his room. He never treated his books as if he possessed/owned them. That is why, perhaps, he preferred to buy second-hand books from the *boipara*. He was also very particular about taking some of his 'already read' books, magazines and other materials back to the *boipara*. The books would come from the stores (where, occasionally, he did buy new books) and stalls; they would live for various times in the study of our home; and then some of them would travel back to the *boipara*. Murmuring and smiling to himself, going through the scribbles and personal notes of others who had possessed the books before him, he would make his own scribbles, and then, after a while, he would visit the *boipara* and sell those same books. If there was one thing that he was very particular about, it was that after some (apparently) indeterminate time, the books had to go, and a newer pile of second-hand books had to come in. In the course of his life to date, most of his books have been replaced, and doubtless the intention is that eventually all of them will be. I was fascinated by the fact that although his attachment to his second-hand books and newspapers was so intimate and personal, it was rarely if ever permanent, and he clearly didn't feel the need to own them materially forever. The chain of exchange – of buying, reading and selling them back to the bookstalls at the *boipara* – had to be continuous. Years later, I too would form this nexus of affective exchange, and thus begin my own relationship with the *boipara*.

Encountering College Street's *boipara*: A public heritage site

The *boipara* is a 1.5 km stretch of College Street in the northern part of the city of Calcutta (now known as Kolkata).[5] This precinct comprises an extensive series of makeshift bookstalls and book shops stretching in different directions. In addition, the area has a cluster of Calcutta's major and small-scale publishing houses; new and established bookstores; a coffee house that has retained the name 'the Albert Hall' since the British era; the prestigious Presidency College (now a University); Calcutta University; Sanskrit College; Hare School; and Hindu School. College Street is justifiably regarded as the educational and literary quarter of the city.

For an extensive period in the past, College Street was a breeding ground for leftist, progressive intellectual life, and consciously ideological, aggressive cultural, social and political practices. Today, it is still known as *the* space where art, culture, theatre, cinema, literature and other forms of creative expression thrive in Calcutta. The ways in which this space has been used completely defy the original purpose of building it. In 1803, the Lottery Committee, which was a town improvement project proposed by Lord Wellesley, had the objective of creating 'an ordered city' (Bandyopadhyay, 2011, p. 85). This plan created a kind of arterial network within the otherwise chaotic and haphazard spatial arrangements of the cityscape. A structured axis was created along and between Cornwallis Street, Wellington Street, College Street and Wellesley Street. However, it transpired that the ways in which the area was planned were completely negated by the ways in which its spaces were used by its inhabitants: College Street is today, and always has been, a 'congested, tramline scarred thoroughfare' (Bose, 2013, p. 219). I suppose in many ways this was a predictable outcome.

While the British were thinking in terms of designing the city from a Western perspective, according to which a beautified, structured and well-policed city reflected and reinforced prosperity and order, the Bengali understanding of a city not only totally departs from the British imaginary structurally but also in terms of language, public behaviour, and interactions in public spaces, as well as in many other minor, less obvious but impactful ways.

While the British divided the city into 'black town' and 'white town',[6] for the Bengali the city has continued to be understood and used as a collection of differing but variously interconnected *paras, tolas, tulis, bagans, ghats*[7] and so on. All these are Bengali words for understanding various localities, which derive their identity partly from the physical structure of an area and partly from the everyday socio-cultural activities that take place in its spaces.

These historical, topographical and architectural tensions between local and colonial/imperial understandings of the city provided an interesting starting point for my thinking about how I might approach reading the spatiality of the *boipara*. Disrupting the attempt at spatial regulation that is colonial College Street is an aggregation of formal book shops, informal bookstalls, formal printing and publishing houses, prominent educational institutions, the historically and politically significant Albert Hall coffee house and numerous shanty tea stalls and other informal eateries. In addition, the area also has a number of stalls and small-scale shops that sell medical goods, stationery and plumbing supplies among a host of other random commodities. Thus, College Street as a whole testifies to the affective coexistence of the *boipara* among other forms of street-based marketing and consumption, and the coexistence of all of these with the formal and informal intellectual and cultural life of the city.

Visually, College Street presents a lively and disorganised impression of residually dignified colonial buildings that house the formal book shops and

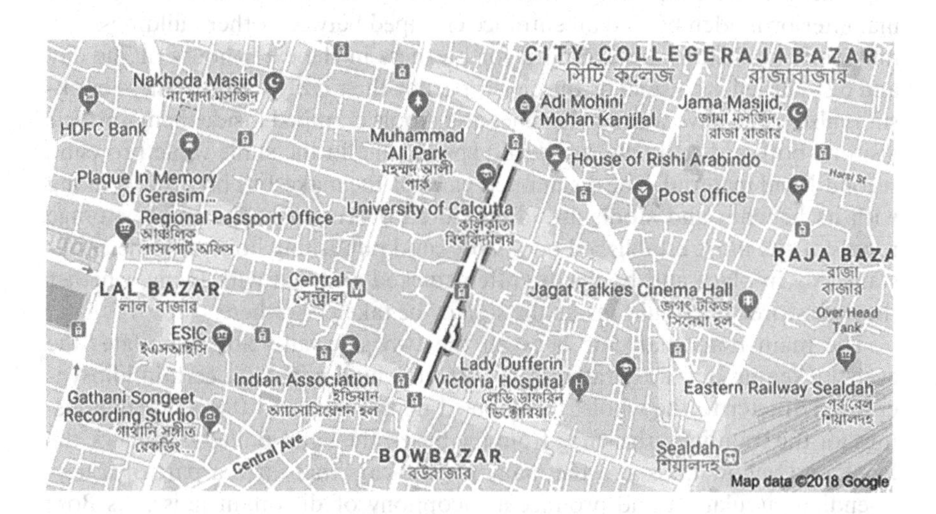

Figure 1.2 A map of the College Street area

Figure 1.3 The stools on the footpath often used by the booksellers, customers and other regulars to sit and spend time in the precinct

publishing houses, including the coffee house, with its vast and striking colonial interior hidden behind an entrance cramped between other buildings. The makeshift bookstalls are arranged for business in the early hours of the day and dismantled at the end of the day as part of their daily routine. These bookstalls also often have wooden benches and stools cramped inside them for their visitors, both old and new, to sit and spend time in the stalls while browsing second-hand material. The dismantled parts of the makeshift bookstalls, along with its benches and stools are often stored overnight in the empty storage rooms of the more established bookstores and publishing houses. Some bookstalls stay in place and books are often stacked and tied with twine, whether they are stored overnight inside the stall or be taken elsewhere.

The main road that cuts between the two sidewalks accommodates the ever-thickening one-way traffic of the street, which imparts a certain idiosyncratic character to the whole: the cobbled main roads and the concrete housings for the tramlines, the smog from the buses, the hand-pulled rickshaws, the black and yellow taxis and the many private cars that form a part of a never-ending circulation and produce a cacophony of dissonant noise. As Bose (2013, p. 220) observes:

> [T]he pavements overflow with makeshift book stalls; half hidden behind them are the established book shops, between which loom the walls of Hindu School, Hare School, Presidency College and Calcutta University. The visible coexistence indicates the symbiotic relation between the educational institutions and the book trade.

This symbiosis indicative of the spatiality of the *boipara* as a neighbourhood is under continual construction and reconstruction. Traversed by innumerable regulars, this space generates and regenerates multiple narratives of everyday life. To experience this space is not a simple matter of decoding it or mapping it; rather, it is of drawing inescapably on it and immersing oneself into it.

Ideas, concepts, roles and relationality

This manuscript engages with concepts and ideas that have proved valuable when 'plugged into' (Deleuze & Guattari, 1987, p. 4) my thinking and/or to the contexts about which I am thinking. Taking my lead from Deleuze and Guattari, I have thus understood concepts as tools, processes or modes of revealing emergences. This has also enabled me to develop and redevelop my understandings of concepts in relation to their *appropriateness for* the contexts in which they are deployed, shaping and reshaping my ways of working with them in relation to the purposes for which I have used them. I have attempted to avoid static frameworks, instead using concepts to make and break problems in order to create 'knowledge of the events'. In writing about College Street and the *boipara*, I have used and explained certain everyday words and phrases with a view to capturing a sense of specificity, not to pin those words down but rather to enable them to retain their mobility. Through my work on this project, I have come to understand that the context opens up the (semiotic/semantic) specificity of words to shifts in meanings so that what emerges is a very particular connection to the aspect of the spatial arrangements in which a particular usage occurs.[8] It is a given that usage and meanings shift in context, but it is crucial to recognise that they do so in very precise ways when the context includes the cultural and personal nuances of lived experience in highly charged and deeply resonant spaces such as the *boipara*.

Within the praxis of heritage, emotion and affect, this trajectory of thinking has been particularly valuable. In examining the everyday spatial movements of a heritage site that is grounded primarily in the material circulation of its components – the books – one cannot help but recognise the complex relationships between affective attachment and intangible heritage (Adey et al., 2014, 2017; Cresswell, 2006, 2010; Urry, 2007; Waterton, 2014) that has aided in the sustenance of this space over decades. Grounded on these theoretical understandings, this book hopes to rethink heritage outside of its conventional static and universal approach and centres around flows, rhythms and movements.

In relation to this, Nigel Thrift's (2008) work on non-representational theory has contributed substantially to the ways in which I think about the relationships between theory, methods and practice. Thrift, drawing on Law (2004), stresses the methodological importance of designing practices that 'are compelled by their own demonstrations and therefore leave room for values like messiness and operators like the mistake, the stumble and the stutter' (Law, in Thrift, 2008, p. 18). I have engaged with 'the ways in which the *boipara* seemed to open itself to my thinking about its spatiality', which has involved engaging with its 'intensities' and 'forces' (Seigworth & Gregg, 2010, p. 1). In navigating the textures of Calcutta in general and the *boipara* in particular, whether informally or purposefully, one encounters many different affective resonances.

Thrift (2008, p. 18) notes that consciousness can be thought of as an 'emergent derivative' of unconsciousness comprising a complex network of 'bodies and things which, knitted together as routinized environments, enable a range of different technologies for more thinking to be constructed'. However, it is the pre-cognitive that prepares our imagination to play in these complex networks. This resonated for me in terms of the way I set out to stress that the everydayness of the *boipara* nevertheless verges on the mythical. The ecology of this space, which thrives on the play of memory and nostalgia, often operates in the ways Thrift describes as pre-cognitive.

Consequently, my spatial engagements notice how things become a part of assemblages – how they can be active binding agents in the production of the 'messy illegibility' of urban space discussed earlier in relation to Highmore's (2005, p. 7) project to 'make legible the effects and affects of illegibility'. While a range of developments related to the independent life and presence of things in our socio-cultural consciousness have been examined, Thrift (2008) takes the discussion further in refusing to separate or create hierarchies between bodies as subjects and things. My conviction is that the human and material must not be studied separately from each other. The strikingly ordered clutter of the *boipara*'s bookstalls and the shared practices of ramshackle building materials, benches, tea urns and so on only intensify the distinctiveness of each stall and stallholder. These few observations – not to mention the extraordinary sense of a mad excess of thingness, the 'spillage' of materiality of every kind across the whole precinct, call out for recognition of the lives of things being completely implicated in the lives, histories and memories of bodies.

Seigworth and Gregg's (2010) work has been useful in several ways, in particular in relation to the in-between and how affect can be considered in terms of 'force or forces of encounter' (2010, p. 2). In the context of the interactions on which I focus in the *boipara*, the circulation of affective intensities – sometimes sticking to bodies, sometimes moving through and in-between the surfaces and relations of human and non-human component parts – becomes vital in the production of the imagined and lived narratives of the space. In relation to his sixth tenet, Thrift (2008, p. 12) writes, 'I want to get in touch with the full range of registers of thought by stressing affect and sensation.' Affect and sensations have been particularly important to my project in ways that go well beyond methodological considerations. What excites me most is that once we are able to realise the primacy and effectivity that affect holds in everyday spatial unfolding(s), while at the same time being completely autonomous, we are freed from concepts of authenticity and adherence. We realise that there is no cultural, political or social 'truth' to our individual activities – there are only honest responses to affect.

College Street's second-hand book market is central to the character of the city because its historical, socio-political and cultural roles have given it a distinctive and unique position in the city's own imaginary, central to its relations with itself. These relations can be tracked, described and attributed, as it is these narratives that have attributed the values of a heritage site to this space. They emerge because of spatial engagements with mundane realities that nevertheless come to verge on the mythic through nostalgic memories, affective resonances, desires and spatial entanglements as productive connections with anticipated futures. To understand the rhythms of College Street, one must understand the space in terms of the connections, processes, networks and matrices involved in the unfolding of a day-to-day reality within a field of difference. The 'College' of College Street refers to the Presidency College (now a university), which was established in the area in 1874 (Bose, 2013, p. 219). In many ways, this original College explains why the area came to have a major coffee house, a cluster of shanty tea stalls, the rest of the schools, colleges and universities, the book shops, publishing and printing businesses and second-hand bookstalls I have described above. This is a space where students, academics, artists and public intellectuals can sit and chat for hours with the tea stall or bookstall's owner or the publishing house proprietors or staff (who may not have had the same level of education). Others merely sit observing the regulars of the coffee house, the students coming and going, the other book browsers and purchasers.

Here, everyday public and spatial practices that are usually guided and regulated by socio-political and cultural tenets, like class and caste – the bearing of one's birth and occupation on one's economic and social status and capacities – in some way melt into relative insignificance: they don't vanish, but they almost imperceptibly coexist. By this I mean that the interactions between the regulars of the space become and remain guided by their affective associations, experiential knowledge, imagination and interactions with the past activities of the space that continually inform the present.

In my most formative years educationally, from the ages of about 17 to 24, I spent a considerable amount of time as a regular of the College Street book market, immersed in the sound of the booksellers shouting out their prices – often including discounts – of books on arts, commerce, science, literature, philosophy, history, politics and so on. Quite early, I began to notice how their voices and words started to form patterns. People walking along that pavement encounter their own versions of such patterns every time they go there, and doubtless they are noticed to varying degrees and evoke varying responses.

For me, the more I was aware of them, the more these auditory rhythms urged interactions in and with the space, opening my senses to the life of the street, to the people, the posters, the books, the buildings, the makeshift stalls, the traffic and so on.

So it was that, when I decided to make this urban precinct the focus of my doctoral project, my consciousness of how aware of these rhythms I had become impacted directly and strongly on how I approached my work. I knew I wanted to explore the complex and multiple temporal interactions that clearly shape College Street's political, intellectual and socio-cultural contribution to the life and character of Calcutta. However, I was resistant to producing a historical account of College Street based in, and viewed through, a descriptive account of its present, using conceptual frameworks relying largely on notions of memory, nostalgia, inter-generational identifications and affiliations. These concepts remain significant, but they are not enough. As will be equally obvious from the approaches I have taken, Lefebvre's (2004) rhythm analysis also proved to be insufficient to achieve the kind of encounter I was seeking with the spatial complexities of College Street.

I wanted to capture the unfolding of the street's spatiality with its multi-layered complexities of time and place, its overlapping rhythms and affects. To this could be added the interplay of the personal and collective, the private and public, yet I became increasingly concerned that these pairings, which could too readily come to operate as binaries, can appear almost meaningless when considered in relation to how everyday life is actually experienced and encountered in the site. As a past regular, I knew that at any given time I was aware of my immersion in the street's dynamics. When I later considered those rhythms that I had discerned, and again thought about them during visits home to Calcutta prior to commencing my doctoral research, I understood that I was actually noticing multiple affective resonances, material and non-material movements, interactions between the past and the present, the human and the non-human, the actual and the mythic, *all at once*. There are no neat bundles to form rhythms or separations: the conjunctions are precisely indicators of conjoining movements – dynamic, active, fluid, interactive. It therefore soon became evident that working in and with this kind of space required approaches to concept, method and practice that could take account of the sensory and cognitive interactions and complexities of the precinct. I needed means to explore how memories and current circumstances collide and coalesce in the apparent chaos of such a setting; how we encounter our own paths and those of others; and how the space itself is animated by and/or responds to those encounters.

This book is a culmination of many realisations, the most important of which is being able to see how the personal is deeply entwined with the spatial. In the chapters that follow, I have attempted to express my understanding of the spatiality of the *boipara*. My writing is neither intended nor enacted as a socio-cultural analysis of its spatial practices, nor do I attempt to provide a comprehensive history of the precinct and the part it has played in the life of Calcutta. My work is better understood as extending on my lived and imagined experiences. My writing entwines my own experiences and knowledge with those of others gathered through my fieldwork process as well as through my literature research and my readings in theory. I also reflexively question the ways in which my writing about the *boipara* informs navigates and revisits my experience of its spatiality. Thus, my book has no beginning and no end. It starts to a significant extent in the middle of things, and I have allowed the stories and rhythms of the space to 'tell me' where the narrative would like to go and thus how the argument develops. This is necessarily informed by the wide range of readings I have undertaken, those whose work has most sparked my interest, what thinking has suggested itself as particularly appropriate to the experiential and affective strands I have tried to capture, the conceptual approaches I have developed and the approaches to scholarly practice I have consequently adopted.

Yearning, anticipation and serendipity, too, make up some of the many strands through which the fabric of the every day of a space such as the *boipara* weaves its multiple dynamics. I say this from experience; although, I am not claiming any particular authenticity or authority. Rather, I aspire to respond as honestly as I can to the affective resonances of the space. In this way, I invite readers to interact with the stories, anecdotes, experiences and ideas as they come to them, deciding for themselves how they would like to reimagine and conceptualise the *boipara*. I do not believe that to think of a space, to imagine it, one needs to be or have been physically a part of it. As I have already suggested, long before I physically began visiting College Street and its bookstalls, my childhood relationship with the spatiality of the *boipara* began when I allowed myself to live imaginatively through the stories, the material objects and the language of the space as I encountered it in my family home, especially through my Dadu and his books. Thus, the primacy lies not in my personal experience of the space, but *in the space*: its stories, its books and other elements that comprise its everyday spatial dynamics. Just as this occurred initially without my presence in the space itself, the same way it can be for readers of this work.

In my research in the existing literature on the *boipara*, I found that apart from a scant number of articles and brief mentions in other contexts here and there (Bandyopadhyay, 2009, 2014; Mazumdar, 2013; Roy, 2014), there has not been much focus at all on accounting for this space in its own right. Over the last few years, general books and edited collections on Calcutta (Chaudhuri, 2013; Gupta, 2014; Ray, 2008) have attempted to capture certain aspects of the College Street book market.[9] While these resources have been helpful in propelling my own thinking about the *boipara*, I recognised two significant things. First, nobody has written expressively and extensively about the unfolding(s) of the *boipara* as a space in the day-to-day life of Calcutta. Most of the articles and books that use College Street at all confine coverage of spatial practices within the limits of historical accounting or reading the precinct as a cultural artefact.[10] Second, the lanes and alleyways, the coffee house and the bookstalls of this precinct have often been used in films and featured in various Bengali novels, by virtue of the spectacular visual experience offered by the precinct. As someone who has been, and continues to be, entangled and emotionally invested in the everyday, insignificant and mundane spatial practices of the *boipara*, I found that these representations of the spatiality of the second-hand book market were too often restricted and fixed, thus never fully liberating the space to breathe and speak for itself.

While there has been some debate within public discourse about whether or not the *boipara* should be recognised as a heritage site, these discussions have never really captured the intangible capacity of the space to 'produce affect' and be 'affected' through the material mobilities of the space. Heritage sites often aim to construct an experience or a sense of place for their visitors. In the case of the *boipara*, the process of this construction is flipped in more than one way. The market never came into existence with a goal of 'creating an atmosphere' (Waterton, 2014, p. 824). A sense of coming together, dispersing and then coming back for more was created over decades in the space – through the books, the booksellers, the vendors, the regular readers of the space, the students, the trams cutting through the main streets and so on. Thus, in many ways, this work seeks to push the boundaries of the current axis between heritage and affect.

It is precisely for this reason that I have been careful not to frame my writing on the *boipara* within any particular discipline or bind it to any specific theoretical or methodological 'framework'. Instead, I have deployed as necessary those different thinkers, philosophers and theorists who have enabled me to liberate the space, drawing on their ideas to help me conceptualise, represent, explore, uncover and interpret the *boipara* in ways that facilitate thinking creatively and reflexively about and with it. As critical geographies of heritage as an emerging field of study move in diverse directions, it is perhaps necessary to approach spaces such as the *boipara*, which undeniably carry historical, social, cultural and political value to the city as a public space, outside of universalist and homogenous narratives of heritage. These narratives often stagnate the experiences of heritage spaces through a lens of 'what happened in the past' and encourage visitors to experience the present through a sense of dwelling on the past. Aiming to deviate from this approach, I hope that through this book we can rethink heritage experience as open ended – an experience that centres the process of the present, of the thrill of being in the middle, not necessarily going back in the past or anticipating the future.

Although I wanted to write about the spatiality of the *boipara* in much more detail than I had come across in any other sources, I have actively resisted this becoming a genealogical or chronological history. Instead, through my writing I set out to capture or witness moments when the personal histories, memories and experiences of the regulars (including me) and the occasional users of College Street intersected with their presents in the space, as well as with what might be thought of as the public, collective or shared memories and significance of the space in the life of the neighbourhood and the city.

It was not the past that I found most interesting; rather, it was the play between the past and the contemporary *in the present* that excited me. I wanted to discover how that which is mobile, in flux and fleeting might be rendered legible affectively as well as epistemologically. As a result, I deliberately fostered a sense of immediacy in the continuing conceptual and methodological thought processes that shaped my primary research questions. What follows is in many ways a collection of interactions, conversations and fragments based in my own experiences and those of others using the space, as well as my responses and theirs to the material, mythic and mundane aspects of the space itself as they revealed themselves through my work. These are not presented randomly because they became the basis for the book itself – that is, for the development of an argument, albeit one that does not set out to reach a particular conclusion.

Notes

1 Ma (mother), Baba (father), Dida (grandmother) and Dadu (grandfather) are Bengali words. Throughout my dissertation, I use Bengali words in referring to these family members, to retain the authenticity of my experience and my relationship with them before I learnt English.

2 The term 'little magazines' carries a similar usage in Indian literary circles to that familiar in Anglo publishing from at least the middle of the first half of the twentieth century. It refers to small periodicals, usually publishing serious, experimental and avant garde writing on literary criticism, philosophy and culture as well as creative work with political and aesthetic import. The Indian little magazine movement originated in many languages, including Bengali, during the 1950s and 1960s. Primarily published and circulated around the College Street area, they have played an important part in Bengali literary and social history. Recognition of the rise of the little magazine movement as a significant literary and political moment in Bengali everyday life is deeply grounded in the spatiality of the *boipara* and its value to the present day.

3 *Adda* is a part of everyday spatial practice in both public and private spaces of Calcutta. It is when a group of friends, relatives, neighbours and others who know each other as acquaintances meet regularly at a certain space for casual conversations. It is an intrinsic part of the coffee house in College Street, but it occurs regularly in many other pockets of public and private space within the *boipara* and elsewhere. I discuss the idea of adda extensively in Chapter 6, in terms of what it means and its importance in relation to the everyday spatial practices of the *boipara*.

4 I have translated this word, based on my knowledge and use of the language, through my everyday lived experience. While Bengali is my mother tongue, I am hesitant to claim any authority for my translation of this word, which has many nuances.

5 In 2001, the name of the city of Calcutta was changed to Kolkata. Buddhadeb Bhattacharjee, the then chief minister of the state of West Bengal, reasoned that this was an effort to do away with the colonial legacy. However, given the complex historical and cultural background of the city, to this day the two names coexist and function in daily conversation, signage and so on. Kolkata is simply the Bengali pronunciation for Calcutta. I use Calcutta when using English and Kolkata when using Bengali. In addition, I have grown up knowing that Calcutta is the name of my city, and from a personal, emotional perspective it makes sense to me to use

Calcutta in a project that explores my personal experiences with a city space with which I have such an intimate relationship. Recognising myself as a decolonised subject, I do not feel obligated to do away with whatever is colonial. The affective, historical and personal association I have with Calcutta does not bow down to any colonial or imperial narrative. Instead, it is deeply rooted in my own spatial and social usages, both in the present and the past.

6 One of the ways in which the British Raj organised Indian cities was through dividing them into black towns and white towns. Calcutta was no exception. The white town was where the British were to live, and the black town was where the Indians were to live. Such a distinction, although very clear and rigid within the rule books of the colonisers, was never successful or effective. In the case of Calcutta, this is because it is essentially an unplanned and accidental city (for more on this, see Chattopadhyay, 2005). Essentially, the boundary between the black town and white town was embedded in colonial thought, but demonstrated virtually no recognition of or impact on everyday spatial practices, because the mobilities – or, more colloquially, the toing and froing between the inhabitants of the two 'towns' were mundane and continuous. The city remained effectively one Indian-and-British city with many neighbourhoods.

7 The *ghats, tolas, tulis, bagans* and so on are Bengali descriptors for local spatial divisions. Most of Calcutta's public spaces, city streets and neighbourhoods have both a Bengali name – which refers to these descriptors – and an English name. As can be seen from the map (Figure 1.1), the British attempts to produce order through the building of wide, straight thoroughfares continues to be undone by the myriad small streets, laneways and alleys between them.

8 This can equally be observed about social, cultural, political and ideological arrangements, but here I understand all these as implicated in spatial arrangements.

9 There are, of course, several newspaper articles that talk about the College Street area in a variety of contexts such as the education system of the city, political movements and other city issues relevant to it.

10 A significant exception to this is the three-volume *College Street er Shottor Bochor* (Ray, 2008) translated in English as *70 Years of College Street*. This recounts a spatial journey through the *boipara* as seen through the lens of publisher, Sabitendranth Ray. Ray worked at the renowned publishing house Mitra and De from 1936, and this series of books chronicles his personal and professional experience with the everyday life of the *boipara*. I happened to read this book a few years before I began this project. I have translated relevant parts from Bengali to English and used stories and anecdotes from his experience as appropriate.

References

Adey, P. (2017). *Mobility*. Routledge.

Adey, P., Bissell, D., Hannam, K., Merriman, P, & Sheller, M. (2014). *The Routledge handbook of mobilities*. Routledge.

Bandhyopadhay, R. (2011). Politics of archiving: Hawkers and pavement dwellers in Calcutta. *Dialectical Anthropology*, *35*(3), 295–316. https://doi.org/10.1007/s10624-010-9199-1

Bandyopadhyay, R. (2009). Hawkers' movement in Kolkata, 1975–2007. *Economic and Political Weekly*, *44*(17), 117–19. www.epw.in/journal/2009/17/notes/hawkers-movement-kolkata-1975-2007.html

Bandyopadhyay, R. (2014, 30 May). The hawkers' question in postcolonial Calcutta. http://dx.doi.org/10.2139/ssrn.2443709

Bose, D. (2013). College Street. In S. Chaudhuri (Ed.), *Calcutta: The living city vol. II: The present and the future*. Oxford University Press.
Chattopadhyay, S. (2005). *Representing Calcutta: Modernity, nationalism and the colonial uncanny*. Routledge.
Chaudhuri, R. (2013). Three poets in search of history: Calcutta, 1752–1859. *Trans-Colonial Modernities in South Asia* (pp. 189–207). Routledge.
Cresswell, T. (2006). *On the move: Mobility in the modern Western world*. Routledge.
Cresswell, T. (2010). Towards a politics of mobility. *Environment and Planning D: Society and Space*, *28*(1). https://doi.org/10.1068/d11407
Deleuze, G., & Guattari, F. (1987). *A thousand plateaus: Capitalism and schizophrenia*. University of Minnesota Press.
Gupta, N. (Ed.) (2014). *Strangely beloved: Writings on Calcutta*. Rupa Publications India.
Highmore, B. (2005). *Cityscapes: Cultural readings in the material and symbolic city*. Macmillan Education.
Law, J. (2004). *After method: Mess in social science research*. Routledge.
Lefebvre, H. (2004 [1992]). *Rhythmanalysis: Space, time and everyday life*. Trans. S. Elden & G. Moore. Continuum.
Mazumdar, A. (2013). Barnaparichoy: A mall in progress, a street in transition. *Subversions*, *1*(1), 122–45.
Ray, S. (2008). *College Street'er Shottor Bochor, vols I, II, III*. Deepshikha Prakashan.
Roy, A. (2014). Reading spaces: Calcutta's *daftaripara*. In N. Gupta (Ed.), *Strangely beloved: Writings on Calcutta*. Rupa Publications.
Seigworth, G.J., & Gregg, M. (2010). *The affect theory reader*. Duke University Press.
Thrift, N. (2008). *Non-representational theory: Space, politics, affect*. Routledge.
Urry, J. (2007). *Mobilities*. Polity Press.
Waterton, E. (2014). A more-than-representational understanding of heritage? The 'past' and the politics of affect. *Geography Compass*, *8*(11), 823–33.

2 The field as it happened to me

The *boipara* I know: The space (I think) I understand

The College Street *boipara* was *the* hotbed of intellectual, political and socio-cultural public discourse when, in the nineteenth century, Bengal experienced what is now known as the Bengal Renaissance. As the social conscience of Calcutta's intellectual class increasingly questioned many Indian orthodoxies, including child marriage, the dowry system, female infanticide and other caste- and religion-based prejudices, some of the most prestigious educational institutions in the city, including the Presidency College (now University) and the Calcutta University, were being established in the *boipara* precinct. Ironically, while Bengal was undergoing a socio-cultural and political revolution based on the adoption of values and aspirations learnt from Britain and Europe, its hunger and determination towards achieving independence from Britain was also gaining momentum. Bengal also underwent the first round of 'partition', which divided the state into two halves: Bengal in the west (with the majority of the population being Hindus) and East Bengal, including Assam (with the majority of the population being Muslims – or, as the British liked to call them, Muhammadans). Lord Curzon, the then viceroy, had ordered such a division in order to create what was clearly a coldly calculated religious division between the people of Bengal that would, he hoped, eventually benefit the British Raj. The aim was to weaken the rapidly growing intellectual and socio-political strength of Bengal. However, this classic colonial 'divide and rule' policy provoked what would eventually seal the fate of the British Raj in India: the emergence of the Swadeshi Movement that demanded the immediate repeal of the partition, claiming that 'the people of Bengal are culturally one and indivisible' (Chatterjee, 1999, p. 112). The students, the intellectuals and the general public of Calcutta made College Street and the *boipara* their refuge as well as turning it into a place for creating strategy. It is surely from these times that the *boipara* became and remains so deeply involved and active in the political and socio-cultural movements of the city and the state.

Both structurally and metaphorically, the *boipara* has played a significant role in most of the political uprisings of the city then and since. Structurally, the labyrinthine layout of the material components of the space created a

DOI: 10.4324/9781003293026-2

perfect environment for having secret meetings and finding shelter during police raids and similar events. For example, there are numerous stories about how the coffee house, the classrooms and the canteens within Presidency College, Medical College and Calcutta University became places where activist academics, young scholars, students and youth in general were given shelter from the political establishment.

There are many stories that circulate in living memories of the booksellers, and older regulars of the *boipara*, about how the book shop and bookstall owners were equally complicit in letting the Naxalites use their space for secret political meetings, plans and protests, as Calcutta became the home city for that movement in the early 1970s. The Naxalite Movement began on 25 May 1967, in a small village in Bengal called Naxalbari. Led by Charu Majumdar and Kanu Sanyal, among others, this movement fought for the rights of landless peasants in Bengal and gradually in the rest of the country. Aligning themselves to the radical left Maoist ideology of China at that time, the movement proposed taking up arms against the government as a part of their struggle. At an alarmingly fast pace, the movement spread from Naxalbari and its surrounding villages to the streets of Calcutta. Before the Naxalite Movement fizzled out in the mid-1970s, Calcutta remained its centre. As occurred during the independence movement, from the nineteenth century to the mid-twentieth century, the students, scholars and the regular intellectuals of the College Street area became active leaders of the Naxalite movement in the city, and the *boipara* became a hub for its clandestine activities. Numerous bookstores and publishing houses, bookstalls, the coffee house, the hostels and canteens of Medical College, Presidency College, Sanskrit College and Calcutta University also experienced random and often illegal raids by government forces.

It was the youth of the universities and colleges in College Street who left their education and future to fight for the landless peasants of their country. Professor Shoma Das, who is now a reader at Nagpur University, recounts in an article in the *Times of India* (Punwani, 2009) that when her father explained to her who the Naxalites were, all she wanted to do was to find Naxals at Presidency University. Years later, she would marry a factory worker and live in a *chawl*. The romanticism of being anti-establishment, education at Presidency College and the seeds of asking questions for the oppressed became a way of life for her. There are countless other instances in urban folklore, anecdotes and popular stories about how College Street came to inspire and support the radical political movements of the city. However, whatever we might try to connect in terms of a continuous lineage of events and significant people in those events, for my purposes what matters is, quite simply, that something has imparted a unique character to the *boipara*. That 'something' is connected to the books – the rich cultivation and circulation of knowledge that has long interacted with the young minds of the students of the surrounding universities, colleges and schools, and sparked their need *to ask questions* about socio-cultural and political realities. In the *boipara*, it seems those realities are inescapably interwoven with myths and memories to such an extent that personal, family, group, community, urban and national histories become as much a part of the myths as the myths become part of how histories are 'recalled'.

I recognised from my first visits to it – perhaps even before – that how the College Street *boipara* functions is grounded in unpredictable, spontaneous interactions, understandings, commitments, and informal contracts and conversations between the various human, material and narrative components that contribute to and constitute the space across time. These alliances and connections are characterised by passion and fuelled by all kinds of affective relations. They are, therefore, best explored using tools for thinking that facilitate the production of non-representational, or at the very least not solely representational, mappings of the space *as assemblage*. This needs to be informed by Deleuze and Guattari's (1987, pp. 12–13) reminders to '*make a map, not a tracing*' and that 'what distinguishes the map from the tracing is that it is entirely oriented toward an experimentation in contact with the real'. That passage stresses that not only a map part of a rhizome, but any rhizome itself is a map.

Deleuze and Guattari (1987) view it as essential to think in terms of connections, links and complications. Concepts thus become tools or processes for the creation of complications, which in turn create 'new connections for thinking' and open new 'planes of thought' (Deleuze, 1995, p. 139). Thinking in terms of rhizomes provides a means to bring these ideas together in the one movement. They explain rhizomatic thinking by distinguishing it from the kinds of 'arborescent' thinking that they see as characteristic of the West, which has focused on locating a central point of a subject of inquiry or investigation from which it branches outwards. This form of thinking has always privileged centres and origins, striving to maintain order, classification and direction. A rhizome, on the other hand, makes 'random, proliferating and de-centred connections' (Colebrook, 2002, p. xxvii). It is crucial to note that there is no intention to produce a binary or opposition between rhizomes and 'tree logic' (Deleuze & Guattari, 1987); rather, they set out to complicate thought by producing a multiplicity. As Colebrook (2002, p. xxviii) explains it, 'You begin with the distinction between rhizomatic and arborescent only to see that all distinctions and hierarchies are active creations, which are in turn capable of further distinctions and articulations.' Rhizomic thinking, then, does not anticipate any particular method, structure or organisation but rather senses a rupture, a gap, and proliferates from there.

Further, once again avoiding the risk of binary oppositions and exclusive disjunctions, we must always put our tracings back on the map. That is, in the case of the *boipara*, what we are able to understand as the 'formal' history of the precinct consists of numerous tracings that inform the map we are making, as do myths, anecdotes and memories, as well as the magazines and newspapers, books, marginalia, scholia and enigmatically embedded love messages participating in the apparently continuous circulation to and from the *boipara* and the bookstalls and stallholders who daily construct it. In these ways, of course, the rhizomes that can be discerned forming and moving within the space, taking off from it and returning to it, connecting to and disconnecting from each other, themselves become part of the assemblages that form between the book shops and bookstalls and coffee house and tea stalls of the *boipara* and College Street and its universities and colleges, and the areas surrounding College Street and the people who move through all of these regularly and occasionally and the city as a whole.

My familiarity with the *boipara* developed long before I had ever actually been there. As described in my Introduction, I started visiting the *boipara* through the books, news magazines, little magazines, periodicals and old notebooks that my family owned and used, and those given especially to me, which, through the hands of my father and grandfather, made many travels between the new book shops and the second-hand bookstalls of College Street and my home. Every year, as I graduated from one grade to another in school, I had to keep my old textbooks, handwritten exercise copies and notebooks[1] in good condition because they were all destined for a particular bookstall in College Street. The old books travelled from home to the second-hand bookstalls and new ones came from the book shops to my study desk. At the beginning of every academic year, along with our report cards, we were handed a list of books along with a number of exercise copies or notebooks required for the coming year. Each school usually has a 'tie-up' or an arrangement with one of the established book shops in the College Street area. In my case, it was Barua Chaudhury & Company, shop number 26. This book shop specialised in selling new school textbooks. Usually, shops like these do not accept second-hand books for purchase, nor do they sell them. The book shops' premises are usually rented ground-floor rooms of ex-colonial residences and other kinds of old buildings that line the footpath. Every year, Baba would take my school list and collect the package of new books, exercise copies and notebooks for each new subject for that grade from Barua Chaudhury & Company. Some established book shops also have in-house printing presses as a part of their business, and these are usually located in the interior lanes and by-lanes that branch off from the main road of College Street. Certain publishing houses also have their own buildings, consisting of printing presses, binding department, editorial section, information and inquiries section and an outlet for their own books at the front of the buildings. The second-hand bookstalls, by comparison, are rather minimal and marginal and operate on the fringes of the pavement opposite the (faded) grand frontages of the book shops in their colonial buildings. Yet the word 'fringes', while it locates them *vis-à-vis* the formal built environment, does not capture the prolific, sprawling growth of the innumerable informal bookstalls that inhabit the space.

Baba and Dadu always stressed that I should read more than just 'curriculum books', so apart from the required books, a host of extra second-hand books – on mathematics, English, history and other subjects – travelled with them from the second-hand bookstalls to me. Then, of course, there was always a collection of storybooks for me from the bookstalls as well. These were, as

Dadu would put it, for developing my reading habits. I would often yearn for more new books, but I was always told that it was the books that mattered, not whether they were new or old. The injunction against new books, though, was never a matter of affordability for my family. It was a matter of choice. Buying second-hand books was important to Dadu, who believed that books needed to be used and reused multiple times. Their contents must be read, re-read and circulated among many people over time; only then was a book fully utilised and valued. Second-hand books, according to him, had much more value and significance than new ones. Every now and then, he described how he loved buying second-hand books over new ones because second-hand books always had many stories attached to them. He loved to wonder who the previous owner of the book was, through the ways in which the book had been handled, the notes in the margins of the pages or sometimes even chits of paper he found inside the covers.

Being in the field: On methods

It is day one of my fieldwork. I am equipped with my notes diary, questionnaire and photographs. I have also made contacts and appointments with a group of bookstall owners and a few other friends from my university days in Calcutta who have agreed to meet me at the coffee house[2] *in the afternoon. I think I am 'prepared'. I have always found preparing for a field trip to be utterly confusing and uncontrollable, especially in this case because I have already experienced the* boipara *so many times. Here is the dilemma that is bothering me while I am walking towards College Street: how do I approach a space that I know and have been a part of from a sense of objectivity that is 'essential' for fieldwork? In other words, how do I observe, study and collect data from a place that I'm so familiar with? At the same time, just because I think I know and have experienced the space multiple times, am I then already imagining and hence taking for granted what might happen the next time? Also, amid all of this where does my preparation fit in? Today I have made a list of bookstall and bookstore owners that I intend to interact with. Some of them I know through the personal time I spent in this space. Some others are going to be new as I have not known them. I will be walking along the footpaths, stopping by for a conversation as and when I see fit. Of course, this is how I have planned to approach my fieldwork while sitting at my study desk at home. As I keep walking towards the boipara from the family home, my doubts about the successful execution of my plan keep increasing.*

The questions I have in mind today are affectual, like the excitement of coming back to this place after a long time, hanging out in the coffee house with my old university friends, spending time inside my own regular bookstores. As the memories rush in, my pace of walking increasing, I remind myself, 'You are on fieldwork. You also must take notes and stick to the plan.' This is the problem with preparation – it always expects you to have a plan and to adhere to it. However, walking along the footpaths towards College Street, observing and listening to the sounds and sights of the streets that lead up to the boipara, *my plan already seems to change. Anyway, I'll continue walking till I reach the* boipara *and see what happens.*

As I cross the main road to enter the precinct of the boipara, *I realise it has been nine months since I was last in India and visited the space. I start walking along the bookstalls. It is around 11.00 am. Some stalls have already begun business as usual; some stalls are still setting themselves up for the day. There are massive piles of books tied with ropes lying all around the footpath. A few customers have come to sell their own second-hand books to these stores. They are waiting for the stalls to be ready so they can start showing their second-hand books and notes.*

Barely a few steps in, I am stopped by Debashsish da. Debashsish da owns Mukta BookMart – well that is what his bookstall is called. He recognises me instantly and asks me to come inside his stall. I have been a regular of his bookstall for almost three years now. I hesitate for a bit. I had planned on going to some other stall in the beginning. However, this is the first natural pause and I decide to go with the flow. We exchange pleasantries; he orders tea for me from the nearby shanty tea stall on the corner of the footpath. The tea seller Poltu recognises me too; 'ki didi bhalo to?' *which translates as 'how are you, it's been a*

Figure 2.1 An everyday early morning scene where the bookstalls and book shops begin to organise themselves for the day

Figure 2.2 The piles of books placed on the footpath, waiting to be organised inside the bookstalls

long time!'. Quickly, not wanting to miss the opportunity, I tell him, 'tomar shathe kotha ache, aschi ektu por' *('We need to have a chat, I am coming to your stall in a bit'). I realise that I have just made my second interview appointment with Poltu. He did feature in my list of people to interact with, but much later on my fieldwork.*

I return to Debashsish da's stall. We begin our conversation. He asks me about my studies in Australia. He starts showing me his new collection of second-hand books that he has just got hold of. We talk about his business; he updates me about a few of my mates from university who are still in Calcutta and are still regular visitors to his stall. Fifteen minutes into the conversation, he points out that I look distracted. I tell him that I am actually doing my fieldwork in the boipara *and I had a few questions prepared that I wanted to ask him. He laughs it off. He replies,*

> *Why do you need questions? You know me, you know the place. Let me just tell you my stories and experiences. I find questions to be interruptive. Plus is my relationship with you so formal that you need to ask me questions and only then I would let you know about the* boipara*?*

It is interesting to note that somehow, he relates questions and answers to a kind of formality that almost takes away from our intimate book buying-and-selling relationship. Excited, and filled with anticipation, I put my piece of paper back into my bag. I continue listening to him attentively. Thus my fieldwork begins at the boipara. *I have not used anything I have prepared so far. I realise, although not in the way I had planned, I have begun to create my own trajectory of interactions for the day. So I continue to move as the* boipara *invites me.*

As these first fieldwork encounters suggest, my experience of the *boipara* is complex, multi-layered, ongoing and inescapably caught up in the intersections between what is known and what is unknown.

My processes of methodological 'design and implementation' (terms I have to regard as provisional given their reference to a space continually under construction) became caught up in the challenges thrown up by the kinds of approaches implied by certain methods. I soon realised that obstacles I confronted when thinking of the 'how' of my fieldwork often pushed me to think creatively in order to find a way around. My diary notes from the first day illustrate the extent to which, once I was 'doing research' in the space itself, my carefully planned methods quickly shifted directions, readjusted themselves and/or required rethinking. It was not that the complexities and specificities of the site had been overlooked in my initial methodological considerations, and I had long recognised the problematics of throwing some kind of preconceived methodological net over a research site. Thus, anticipating a range of challenges, I had allowed for the importance of mixed methods, flexibility, reflexivity and responsiveness. However, 'the field' almost immediately responded to me in ways I had not anticipated, despite my intimate insider knowledge of the research site. There was very little likelihood of productive outcomes from any 'ready-made' approach to research design in this project. So, instead of the kinds of questions that one routinely asks oneself in implementing reflexive approaches to methodological design, I needed to navigate towards more effective processes by asking myself more precise questions regarding how I might develop methods even more relevant to working in this unique site and able to support my specific project of undertaking a spatialisation of it.

Tim Ingold (2015, p. viii) notes that methodology should always be about 'trying things out and seeing what happens'. Writing was, of necessity, my principal means of producing this book, but having recognised that the *boipara* was even more complex as a research site than I had imagined, I became concerned that writing carried with it a danger of 'pinning things down'. In writing *about* the field I could lose the sense of the mobile, volatile and ongoing experience of *being in* the field. I have therefore understood my writing about the *boipara* as an extension of being there. In doing the writing that has been impelled by my experiences of moving through the *boipara*, by my planned and unplanned encounters with people, things, affects-and-events, I am in a sense entering into a process of 'doing the *boipara*'. That is, I perform my experiences and conversations, my observations, scholarly reflections, connections, ideas and conceptual manoeuvres as if they are co-participants in the enterprise.

Interdisciplinary theory–concepts–methods

Methodological choices in recent research in cultural studies, cultural geography and studies of spatiality demonstrate an increasing recognition of the need to diversify approaches and deploy mixed methods in order to address the complex demands of interdisciplinary research. In responding to the need to challenge not only theoretical frameworks and ideas but also the methodological approaches that help us produce and develop them, we can recognise a shift in emphasis – in cultural geographies and other allied studies of culture and society – according to which,

> [O]bjects of enquiry have become more fluid – focused more often on practiced, ongoing life than cultural artefacts – seeking less to tidily encapsulate things than to show how things always exceed their concepts, and how the world is inevitably messier than our theories of it.
>
> (Shaw, DeLyser, & Crang, 2015, p. 212)

My primary focus is on thinking through the personal and the spatial, and I do not classify my project within the boundaries of any discipline. Rather, I use ideas from a range of fields as and when they are pertinent and useful to my processes. This interdisciplinarity allows me to think of my work as taking place, accordingly, in a liminal space, by which I mean an *in-between* that in various ways trails into and out of what is around it.[3] My approach to researching the *boipara* as a space involves working in-between, in which it involves conjoining my experiential knowledge *and* that of the other regulars of the space *and* scholarly perspectives.

As the various components of the space organise, dismantle and reorganise themselves through their daily spatial practices, our material, affective and sensorial registers of the space change accordingly. Thus, in perceiving the same space with slight differences every time we encounter it, we create and recreate stories and experiences. Using the concept of the assemblage, I invite understanding of the *boipara* through the numerous links, connections and complications that are formed as a result of the continually changing actualities and perceptions of the space.

Identifying and explaining my key epistemological and conceptual influences, their usefulness and various problems associated with them emerged as a valuable process. No matter how we approach space as everyday citizens and researchers, we are always 'doing space' along with how it is 'doing' itself. Further, thinking is doing and doing is always informed by thinking. When we engage with the field, we can too easily fall into a trap of orienting ourselves *either* to a conceptual framework *or* to a methodological trajectory we have already outlined. I have tried to move away from this. I wanted to maintain an 'experimentalist orientation' (Sheller, 2014) towards my work in the space, at least partly because I had become alert to *how* we 'do' space while we are also implicated in how it is 'doing' itself.

As a result of the initial experience described in the diary entry that opens this chapter, when it became very clear that, no matter how thoroughly I had tried to think through my mixed methods, the space and its users were not going to be passive participants in my data collection. I determined to resist allowing my theories, concepts and methods to dictate how I engaged with the space and the trajectories taken by my engagement. Thus, when I thought about or with the *boipara*, it was as much through its stories, its interactions among its components and its everyday predictable and unpredictable movements and revelations as it was through my experiences of it. The field is *in itself* the fabric of the *boipara*, one version of which this book produces through multiple stories. The *boipara* is not some kind of container that holds memories, experiences or stories waiting for a researcher to find them. As a consequence, in a double movement between ontology and epistemology, I set out to work *with* the space by trying to make sense of it *on the move*.

The *boipara* and the various forms of agency involved in producing the data are no longer limited to the human but invoke all other components and aspects of the relational network of assemblages (Braidotti, 2012; Coole & Frost, 2010). The motive in research thus becomes to make multiplicities (Deleuze & Guattari, 1987) – that is, to see where complexity can take us towards thinking productively, even though the aim of research has often been the opposite: to render the complex as simply and straightforwardly as possible.

I have been coming to the boipara *for six days in a row now. My everyday interactions so far have been nothing like I had planned. The other day (Day 1), I bumped into Debashsish da whom I had known for years. I notice how my fieldwork is moving forward and backwards in different directions and layers. However, the movement is completely different to the ways in which I had envisaged my fieldwork unfolding. I thought I had known the space, used it for so many years and hence my plan to navigate the space would completely work out in the ways in which I had 'planned' it. I thought I would walk into the* boipara *and start talking to the first stall on the first part of the footpath and continue from there. However, Debashsish da happened to me! I interacted with him, and then interacted with some of his friends, to whom he introduced me at the end of our interaction. I have been following his trajectory through the* boipara *for the last three days. But, wait, the tea stall owner, Poltu, referred me to Basak Book Store, in between. Thus, it was a halt, a pause in the trajectory I was following. My interaction with Poltu created another entryway.*

Basak Book Store is a big two-storey bookstore inside one of the laneways that branches out from the main road of College Street. They have a printing press and publishing house on the ground floor of their massive store. I am meeting the owner and the manager for an interview today. Poltu has supplied tea, about four times a day, to the workers of this store for the last 15 years. He claims he knows inside out of the store and the printing press room. His description of the routine of this bookstore and those who work here is vivid and detailed. I had made a note of his observations before. I am going to see it for myself today.

There is always someone who claims they know the routine and the everyday movement of the boipara *or even the stalls and stores individually. Poltu thinks he knows when the manager of Basak Book Store comes; he has an idea of the time when the store is most busy. He pointed out to me how it was in the late afternoons or in the evenings the young writers and poets would come to the store, pitching their ideas to the editor, Asitbabu, and the owner, Tarunbabu. Poltu warned me that this was probably not a good time to go as they discuss serious matters then. Poltu, sitting diagonally on the corner of the footpath across from Basak Book Store, had a keen eye for what goes on inside the store. Even though he only went three or four times daily inside the store to serve them tea, sitting outside he nevertheless knew very well who came in and who went out.*

I just spent around an hour in Basak Book Store. Poltu's descriptions were quite accurate. The coming in and going out of the customers, the manager and the owner going through inventories and speaking to suppliers all happens later in the morning, as Poltu had told me beforehand. The manager, describing his daily routine, said the same things while giving me a tour of the bookstore and the printing press. However, here is where Poltu was mistaken. Contrary to his warning that late afternoons are when the editor and the owner of the bookstore are usually busy listening to potential publishing pitches and hence should not be disturbed, I was invited by Asitbabu, the editor, to sit in their sessions today. Asitbabu, now aware of my educational background, thought my inputs to some of the pitches would be valuable.

[In relation to the staff of Basak Book Store] … all this while I thought I was an outsider to their everyday routine and their space and that Poltu knew it all. However, now I was suddenly given entry to a part of their everyday routine that Poltu knew was inaccessible to him and thought would be the same in my case. Am I suddenly now holding a position of privilege (informed through class and education) within this everyday routine? Am I suddenly an insider in a routine that I had no idea existed? Poltu thinks he knows and understands the minute details of the routine of this bookstore. I got an entrée into these movements through him. Now I am a part of an important segment of this routine that he has no access to. [At first I found myself wondering] … whose experience is more valid then? Is Poltu the insider or am I the insider? Yet both of our experiences are real, as he lives through this routine every day, and I had my share today. However, my understanding of my position in the field slowly blurs.

My experiences in the field and my understanding of those experiences increasingly moved between various degrees of intimacy and ambiguity. Here, Poltu the tea stall owner had obviously made up his own mind about my unsuitability to participate in 'serious' intellectual business. I can't know whether this was to do with my gender; his unfamiliarity with what is indicated about my education by the words 'fieldwork' in the consent form he signed; the simple assumption on his part that he is more insider than me; or none or all of these. We might readily observe that his advice that I should not go to the store at a time when 'serious matters' are discussed speaks of his level of cultural capital rather than mine, but as to how it positions him in relation to me or me in relation to the site of his daily work of which he possesses much more intimate knowledge than I ever can is a very different matter. As a reflexive researcher, my confusion about all of this is revealed in my diary notes, but what gradually became clearer is the extent of my personal and ideological dilemma: the more I worried about these complexities, the more things seemed to 'blur'.

However, I also note that these dilemmas were thrown up in process and by practice. They are further incidences of how the site could so readily undo my sense of what I thought I was doing in researching it; what I thought I already knew due to what I felt was lengthy and intimate previous experience; and how I was rapidly recognising that what I already understood was much less than I believed. When I entered and interacted with the *boipara* from a new perspective as a researcher doing fieldwork, the extent of my 'insider-ness' was revealed as less clear-cut yet more impactful than I had anticipated.

The spatial movements of the *boipara* that I think I do 'know' and 'understand' are deeply personal, intimate, unique and ambiguous. This means that where I stand in order to start thinking about this space is right in the middle of things. I already understood that there was no 'objective' picture of the *boipara* to be painted, no matter how detailed a combination of sociological, cultural and political analysis might be applied. I appreciated that there was a far more complicated process at work, and I tried to set my work up in ways that could deal with that. However, what my work in the field allowed me to understand is that the field itself is a far more active participant than I had even begun to anticipate.

All this underpins the choices I have made to 'go with the flow' of the fieldwork, how I have structured and written this book, and the voice(s) I have tried to find in my writing. The continual conversations between the complexities of affect, senses, ideas, values and the modalities of varied threads of narratives (across different moments) must be acknowledged even as I recognise that they cannot be 'really' captured. In other words, it is only when I try to explain how 'I think I know the space' that I can begin to discuss 'how I have come to understand it'.

Listening to the field: Noticing affective circulations

I have found the idea of affective resonances, which I first encountered in Ben Anderson's (2008) work, a particularly exciting way of capturing how a space is activated by the movements, gestures, reactions and expressions that people use when they talk about their experiences of and in that space. In my work, these people are the regular users of the *boipara*, who participated in my project directly or indirectly. Affective resonances are always 'associative and collective' (Anderson, 2008, p. 146). This means the regulars with whom I talked have not autonomously come to realise these resonances in the process of describing them. Functioning within the complex site of the *boipara*, they interact with existing material aspects of the site, associated narratives and other people who use the space, so each participant becomes a catalyst for affective resonances to emerge through their expression of experiences and memories and so on, but those expressions already capture the experiences and memories of others. The *boipara* thus becomes a 'being together of existences' (Amin & Thrift, 2002, p. 28). This being-togetherness, and its associated interactions and ongoing transformations, create circumstances for the production of affective resonances.

We can note that, as with stringed instruments or any number of other material examples, resonances are vibrations that pass from one body to another or from one part of a body to another, while amplifying themselves in the process. This offers something of an analogy for how the interactions between space, time and bodies take place in a site such as the *boipara*. At any given moment, and in any given situation and conversation, there are historical, cultural, political and sociological narratives of the *boipara* from different temporalities informing the present and producing conditions that will produce future emergences. Simultaneously, a continuous intersection of past affects with those experienced in the present as taking place in and through the spatial movements of the *boipara* also contributes to how the place/space is sensorially activated and expressed. Affects cannot be differentiated from the space–time narratives nurtured by material and bodily encounters. In these ways, the unfolding of the *boipara* with its multiple, relational experiential knowledges and the trajectories of manifold histories continuously enmeshed with the present is simultaneously 'immense' and 'intimate' (Massey, 2005). The *boipara* is eventful at any given point in time, but also continuously becoming-otherwise. It would be impossible to write a history of the spatiality of the *boipara* in a neat teleological, linear narrative, since this would inevitably produce an exclusionary project, privileging some stories over others. Similarly, one cannot expect to have settled pre-planning of how the field should be studied for all the reasons I have explored in this section.

The complex environment of the *boipara* starts to make random, proliferating connections even before we enter it. These connections, disconnections and ruptures can never be controlled, and it is exactly for this reason that, when appropriate, they should become cues for flexible rearrangements of research methods. Noticing such moments opens out processes of 'studying the field' to the creative, unpredictable and eventful. I was using photo-elicitation to prompt conversation with a bookstall owner when our discussion was interrupted by his prolonged engagement with one of his regular customers. As it transpired, the interaction that followed between the bookseller and his customer provided invaluable material for my research. An apparent 'disruption' to my methodology also revealed a range of other ways in which I might think about the everyday spatial movements of the *boipara*, and thus observational research and casual conversation about what I had just witnessed took over from photo-elicitation.

Massey's (2005) second proposition helps explain my position by emphasising multiplicity and plurality. Positioning myself as a researcher has continually involved complexity, especially given that my responses to the *boipara* have helped frame my research rather than the reverse. My experiences of the *boipara* have involved different temporalities and spatialities from multiple sources. Any attempt to establish a stable position from which to think about the *boipara* is clearly problematic because not only is the space itself complex and fluid, but also there is no fixed and absolute position outside any social space from which to observe it. This is exemplified by my work on the daily routine of Basak Book Store, during which I found myself constantly in the middle of the insider–outsider tension. From Poltu's position, he was the insider in relation to the everyday routine, informing me of the accessible and inaccessible times to visit the store. However, when I lived the same routine, I became an insider to something to which Poltu probably never gains access.[4] It is perhaps not hard to determine that Poltu does not have formal education, and hence is unlikely ever to be invited to an editorial meeting in a publishing house. However, there is also no question that Poltu's knowledge of the routines, movements and general nature of the space is much deeper than mine. Thus, between us we shared moments in an everyday routine that was sometimes informed by common experiences and sometimes separated by our differing socio-cultural positions. It is evident that we differ in terms of degrees of individual cultural capital in relation to this public space at any given point in time, yet by experiencing a part of a daily routine, both collectively and individually, I realised that our everyday socio-cultural positions are also in constant movement and negotiation with each other. These movements continually destabilise our respective positions in the field.

Having had a close, multi-layered affective and physical relationship with the space, I identify myself as an insider. However, writing this over time – sometimes in close proximity to the *boipara* and sometimes from a great distance – of necessity there are aspects that I have not captured of the ongoing unfolding(s) of the space. I am, then, an outsider to those moments and events. The in-betweenness that comes with being an insider and outsider was further complicated during my fieldwork by, for example, material conditions in the street or changes in the weather. There were days when the bookstalls undertook their daily work with slight changes. On a rainy day, for instance, some booksellers reorganised their stalls in ways that would protect them from getting wet, rearranging their books and putting out sheets or tarpaulins from the outer edges of their stalls. Used, reused, handled and mishandled repeatedly, the pages of second-hand books frequently have a fragile texture: a drop of water could tear the pages or smudge the print. These were important things for the booksellers to consider. Their makeshift, wet-weather protection arrangements would also serve as pedestrian stopovers, providing shelter from the rain for passers-by. It is absolutely commonplace that someone walking in any city will stop under any shelter they can find and wait for the rain to subside. In my fieldwork, though, I noticed that the impromptu reorganisation of the books, the putting up of the sheets and the stopping of pedestrians, creates scope for a different atmosphere and thus a different experience of the place.

This is even more noticeable if one visits the same bookstalls and walks along the same footpath regularly, as I did for my fieldwork. When it rained, how I had come to know the space through my repetitive, routinised behaviours was changed in a matter of minutes. When my research practices – which had become habits that intensified my sense of insider-ness – were transformed by a change in the spatial circumstances, I felt like more of an outsider. Simple experiences of this kind can invite one to wonder whether the research practices one has adopted enable one to find the right kinds of questions to ask about the field and about one's position in relation to it. In such ways, I quickly came to understand – with some force – that research as a practice needs to be as open to transformations as the field in which it takes place.

All this points to the fact that one can never develop full grasp of one's experiences, affective responses and continual weaving of narratives within and about a space like the *boipara*, which is so strongly characterised by interrelations and transformations. If we embrace a sense of our ongoing vulnerability, naivety and *not* knowing, it can liberate us from asking questions such as, 'Who speaks for the space of the *boipara*?' I say: everyone. Everyone who converses within the *boipara* also converses about it. As both an insider and an outsider, and a person moving continually between those positions, I am a multiplicity and at the same time I become part of the assembling of the space.

The more we problematise the notion of positioning oneself in relation to a research site, the more we are likely to experiment with responsive and open research practices and multiple positions, and thus the more we can liberate ourselves from the individualistic obligation of conventional research methodologies. Sheller (2014) notes that in recent times, more and more disciplines have been rubbing their methodological boundaries against each other to work experimentally and collaboratively in developing newer forms of research and knowledge creation. For example, Sheller notes that Hayden Lorimer, a cultural geographer, works with artists, while Mike Crang uses photography extensively. Trevor Paglen is a practising geographer and artist, as is Tim Cresswell, from whom I have drawn significantly in relation to his politics of writing, and who is a geographer as well as a poet (Sheller, 2014, p. 133). Collaborations and transdisciplinary practices provoke changes in methodological thinking, which can free the researcher from the obligation to draw data *from* the field. Instead, one can focus more on being in the field, experiencing and navigating through its complexities, concentrating on 'knowledge creation' rather than knowledge analysis.

In-between methods

Does our engagement with the field start when we enter it and end when we leave it? My diary notes from Day 1 demonstrate how thinking about the *boipara* while walking towards it impacted my pre-planned fieldwork design. I began noticing the multiple movements of the space before I even reached it and was already changing my methods before I had a chance to implement my carefully sequenced interviews. Is it any more likely that I would simply switch off from engaging with the *boipara* the moment I was out of the field? Since *thinking about* the *boipara* is actively engaging with it, often blurring boundaries between imagined, anticipated and lived experiences, it is essential to acknowledge that thought about the field is part of my methodology. Whether thinking about the *boipara* occurs while I am walking towards it, while I am in it or after I leave it, I cannot ignore such thoughts as things that somehow happen 'outside the field'. They are thought practices and processes that inform my experience of the space, and thus my modes of responding to it and writing about it.

Conventional modes of framing and implementing qualitative methodologies remain as useful as they ever were, but they are also transforming and interacting as they encounter different questions, newer problems, and spatio-temporal complexities that push their limits and force us to think creatively and productively about how we understand research methodologies. Nigel Thrift (2000, p. 1) points out that researchers in the field of cultural geography have allied themselves with a number of qualitative methods, most notably in-depth interviews and ethnographic "procedure". Going to the heart of my concerns, he notes that

> what is surprising is how narrow this range of skills still is, how wedded [cultural geographers] still are to the notion of bringing back the "data" and then re-presenting it (nicely packaged up as a few supposedly illustrative quotations), and the narrow range of sensate life they register.
>
> (Thrift, 2000, p. 6)

Thrift encourages us towards a methodological paradigm in which 'our theoretical talk can be used not only to interrogate established methodologies' but also to push the boundaries of the methodological strategies themselves (Thrift, 2000, pp. 1–6). It is clear that the changes for which he called more than two decades ago are taking place. A project like mine that tries to look at a space as a system of emergence must think of methodology as a tool that problematises or diversifies data instead of collecting it. Better still, I need to think of methods as 'entryways' (Manning & Massumi, 2014). However, for these kinds of shifts to happen, I must acknowledge my own implication and transformations within the spatial movements of the *boipara* as a multiplicity. A multiplicity is the 'effect of its connections (or becoming-multiple)' (Colebrook, 2002, p. xxvi). Bonta and Protevi (2004, p. 117) note that a multiplicity is defined 'by its lines of flight or thresholds where qualitative change in the system will occur'. It is useful to understand my own position in the *boipara* through these recognitions.

Having experienced the space through intimate narratives, anecdotes and experiences of regulars, such as my father and grandfather long before I had physically become a regular myself, I began with the objective of framing my research methods in ways that enabled me to test whether my 'analysis' could resist the urge simply to align with the discursive frameworks already provided by historical analyses and popular cultural narratives and also to explore how a space can operate and be understood in relation to the movements involved in assemblages.

Photo-elicitation

Photo-elicitation is a form of qualitative visual research in which photographs are used during research interviews in order to elicit richer, often more narrative responses from interviewees. John Collier, an anthropologist who was one of the first to use the method, explains that it is capable of opening up 'emotional revelations' and 'psychological explosions and powerful statements of value' that are almost always missed in conventional data collection processes (Collier, 2007, p. 62). Sarah Pink (2007, p. 68) points out that photo-elicitation 'implies using photography to elicit responses from informants, to "draw out" or "evoke" an admission or a social fact from a subject'. As a practitioner-researcher, Pink has become a prominent advocate for the method, arguing for its integration 'with the new ethnography' through redefining it as 'a mode of collaboration in research' (Pink, 2001, p. 68). For me, the significance of such a way of experimenting with the potential of the interaction between researcher and participant is precisely that a photograph can function as an opportunity to enrich and complicate the exchange through the sharing of memories and of stories, elaborating on the 'moment' in frame, the provision of background context and, in all likelihood, details about people, the material environment, object relations and so on. All of these add to the desire to 'place' the scene in relation to the 'real life' of the subjects, the site and the time. This produces a more conversational exchange, a sense of collaboration between researcher and participant in 'filling out' the photograph, which is far more difficult to achieve in a conventional interview, no matter how 'unstructured'. A photograph, whether the participant or the researcher has taken or provided it, whether it is current or historical, introduces an opportunity for shared evocation of meanings. Thus, the visual 'data' in a framed image is capable of provoking responses that go beyond singular temporal or spatial subjective memories or imaginaries.

The photographs functioned in many ways to enable a kind of departure or escape into unknown and unpredictable narratives. For example, in one situation, a *boipara* regular talked about his experiences by primarily reacting to some of the physical features of the pictures, like colour, texture and the quality of the images, rather than the content. In another case, a picture of the interior of one of the bookstalls triggered a story of times spent inside the coffee house. The coffee house was not visible in the picture, but for the respondent it was nonetheless evoked by association with certain details in the photograph.

There are two things to think about here. First, photographs reintroduced to a spatial assemblage do not function simply as agents or catalysts producing data. They become active in prompting recollections of events, objects or places that are not directly portrayed. They can also change the relationship between the researcher and the subject of the research in the sense that through conversation these two people (me and the person being shown the photographs) together weave their individual experiences into relationships that point beyond the limits of either's personal responses. Some people who were reacting to the photographs expressed feelings of anticipation (e.g. predicting the future shapes the *boipara* might take) based on their experiential knowledge. At the same time, my own experiences of the space, as well as my role as researcher, meant that my reactions often resonated with their stories, whether or not I raised that with them. These complex interactions invite one to problematise whether such research is primarily eliciting or extracting data from photographs, and what the impacts may be of the researcher being drawn into participating in those responses – overtly or implicitly – with both predictable and unpredictable affective consequences.

Semi-structured interviews, conversations and observational research

Hesse-Biber and Leavy (2006, p. 125) note that 'semi-structured interviews rely on a certain set of questions and try to guide the conversation to remain, more loosely, on those questions'. While it is quite liberating to be in a space that inherently functions through a logic of relationality and continuous construction, for me as a researcher, there was sometimes the lingering fear of not having any control over what I was about to experience as a researcher. Thus, I had prepared a series of questions which I hoped would function primarily as trigger points for conversations that would take us beyond the scope of the question itself.

The semi-structured interviews were designed mainly as entry points or prompts for further stories, anecdotes and memories. Although the term 'semi-structured' is used, there is too often a preconceived expectation with interviews that the response will paint a coherent picture of experience. Experiences never function in isolation and autonomously. In the case of the *boipara*, the responses to the semi-structured interviews were always context, environment and atmosphere dependent. Each was, in other words, 'an encounter in its own right, chock-full of awkward pinch points, backward tracing realizations, and cascading memories, replete with subtle transitions, over brimming with heart-thumping intensities and felt emotions' (Bissell, 2018, p. xxxv).

Thus, even if we think of responses to semi-structured questions as data, what counts as data is never restricted only to words or, separately, to other kinds of reaction. Nor do those responses address directly the questions asked, but proceed instead along lines of association. For example, on numerous occasions booksellers who were asked about their 'everyday routine' recounted those in detail. Their precise and thorough descriptions, which combined both memory and contemporary life, *expressed* their emotional attachments to the space, something they may not have been able to describe directly.

Their emotional intensities varied when talking about different aspects of the *boipara*. Some spoke with a deep nostalgia for the past that they seemed to miss today. Some were positive, hopeful, even joyful about their prospects for the future and that of the *boipara*. Some were happy and proud that I was writing about the *boipara*, while others questioned the purpose and point of my work. These reactions, combined with their stories and anecdotes from their routine lives, were all part of the complex set of relations they each held with the *boipara* and it with them. This was not merely incidental information and, from a methodological point of view, this underlined the fact that observations and experiences can never be separated from interviews and interactions. At the same time, researchers must be reflexive about the fact that, while witnessing data unfold, we become active participants in its production even as we 'collect' it.

Thus, while trying to attend to my interviewee and what he might be doing while responding to my questions, I was also often distracted by the movements and behaviours of the students who had come to buy books. They looked at me and I looked at them. Some of them asked me about my project, starting up a conversation that would go in its own direction. After a few minutes of these interruptions, I would have to turn my attention back to the bookseller and try to pick up the threads of the previous conversation where we had left off. While all this was happening, other exchanges (e.g. agreements and disagreements between the students and the bookstall owners) were also a part of the unfolding spatiality for me, and hence important to observe. Similarly, at times some of the ways in which the students behaved and interacted with each other, looking through the books, reminded me of my own time here when I was a student; although while there were many points of familiarity, there were also some noticeable differences. I made notes of 'as much as I could', but there was always more. In these ways, observational research is effectively inseparable from other forms of qualitative research undertaken in the field.

Autoethnography

Autoethnography 'allows us to use our own experiences, thoughts, feelings and emotions as data to help us understand the social world' (Hesse-Biber & Leavy, 2006, p. 184). Because of my close personal familiarity with the everyday environment of the *boipara*, I chose to take up a critical and reflexive autoethnographic approach to thinking about the space, being comfortable about using my own experiences and personal thoughts. Experiences, memories and acts of remembering are not frozen in time and static in space; they move and rearrange themselves to present different versions of the same experience every time we revisit them. This is mainly because of the play between the present experiencing self and the past as one engages with and experiences that past at any particular time in any particular place in the present. Since the present is always mobile and changing, the dialogues and negotiations that the present and the present experiencing-self have with the past are also mobile and changing. My use of autoethnography has enabled this sort of play between past and present, since it has allowed me to use my own experiences at different times: from when I first heard about the *boipara* to my many visits over the years, as a visitor, as a student and as a researcher, and later when I revisited my experiences while thinking and writing about the space when not physically *there*.

Comments about a rally from the bookseller, Rashida

> *I do not think this is ever going to be possible. College Street from time immemorial has been known to be the space where students come to voice their opinions through organised rallies, protests and marches. To take away the right is to take away their right to this space. In addition, students come to colleges, universities, to become political beings. It is only their right that they get to use this space to express their voice. Didi, if there is a fee hike in one of these colleges or universities tomorrow, where will the students go to protest about it? The protest against student fee hikes, or student seat reservations during admission process, or any other university administration or student-related issue should be raised and negotiated at the space that is designated for them – the campus grounds, the boipara. Tell me Didi, what is more important, the noise or the right of the students?*
>
> *(Day 11: Conversations with Rashida)*

These remarks, prompted by the rally outside, disrupted the flow of our conversation and I could thus have treated them as nothing more than a momentary distraction from the flow of my focused research. Accordingly, I suggest that an effective methodology must allow for these in-between moments, pausing to notice, register and, then or later, reflect upon the eventful everyday rhythm.

Notes

1 Materially, a child's exercise copies and notebooks were of value to greengrocers and shopkeepers because they made paper bags from them. However, effectively the travels that these used papers went on were infinite. They had many other possible lives – including, most importantly, how they became tutoring and learning material for other children who could not afford books. That is, children would study the work of other children as a means of self-education and/or enhancement of their basic education.

2 As mentioned earlier, the formal name of the coffee house in College Street is the Albert Hall. However, this name is hardly used within the everyday language. The place is always referred as 'College Street's coffee house' or simply the 'coffee house'. As a result, I have chosen to call it 'the coffee house' throughout the book.

3 That is, I am not referring to 'liminal' as a relatively dramatic threshold moment, which is a common conception among, for example, people thinking about certain psychological states, rites of passage, initiations or mystical/shamanistic experiences. The notion of the threshold as used in architecture is perhaps more

apt – where the threshold of the house functions as an everyday space between the street and the house, between public space and private space. However, the liminality to which I refer relates more to a biological, ecological and geographical usage, in which the word evokes the ways in which different parts of a landscape meet in spaces that contain some aspects of soil features, plants and animals that trail into it from both parts, whereas they appear in the in-between from settled spaces, spaces that can be named as this *or* that kind of environment or landscape. This usage is obviously similar to 'littoral zone', but that term refers to the coastal space between high and low tidelines as a specific environment inhabited by creatures specifically adapted for and located in that zone. In a liminal space – say, where the beach trails into the dunes and the dunes trail onto the beach, or where aspects of coastal scrub trail into rainforest, and vice versa – particular creatures and plants are specific to what is on either side of the liminal space, while the liminal space is where they brush up against each other from both directions. When they do meet, they are always in-between.

4 I am aware of the sociological implications of this interaction. I was readily invited to the editorial meeting that Poltu thought was a no-go zone. My educational background and social class gave me access to this part of the store's daily life from which Poltu was excluded. However, I want to draw attention to something more complex, which is that while we are both creating our realities through the same narrative in the same space and time, his differs from mine in certain precise ways.

References

Amin, A., & Thrift, N. (2002). *Cities: Reimagining the urban*. Polity Press.

Anderson, B. (2008). Entries for affect: Emotional geographies, non-representational theory. In D. Gregory, R. Johnston, G. Pratt, M. Watts, & S. Whattmore (Eds.), *The dictionary of human geography* (5th ed.). Wiley Blackwell.

Bissell, D. (2018). *Transit life: How commuting is transforming our cities*. MIT Press.

Bonta, M., & Protevi, J. (2004). *Deleuze and geophilosophy: A guide and glossary*. Edinburgh University Press.

Braidotti, R. (2012). *Nomadic theory*. Columbia University Press.

Chatterjee, Partha. (1999). Modernity, democracy and a political negotiation of death. *South Asia Research*, *19*(2), 103–19.

Colebrook, C. (2002). *Gilles Deleuze*. Routledge.

Collier, P. (2007). *The bottom billion*. Oxford University Press.

Coole, D., & Frost, S. (2010). *New materialisms: Ontology, agency, and politics*. Duke University Press.

Deleuze, G. (1995). *Negotiations*. Columbia University Press.

Deleuze, Gilles. (1987). G. Deleuze, F. Guattari, A thousand plateaus–capitalism and schizophrenia 2, trans. B. Massumi.

Deleuze, G., & Guattari, F. (1987). *A thousand plateaus: Capitalism and schizophrenia*. University of Minnesota Press.

Hesse-Biber, S. N., & Leavy, P. (2006). *The practice of qualitative research*. Sage.

Ingold, T. (2015). *The life of lines: A brief history*. Routledge.

Manning, E., & Massumi, B. (2014). *Caught in the act: Passages in the ecology of experience*. University of Minnesota Press.

Massey, D. (2005). *For space*. Sage.

Pink, S. (2001). More visualising, more methodologies: On video, reflexivity and qualitative research. *The Sociological Review*, *49*(4), 586–99.

Pink, S. (2007). *Doing visual ethnography: Images, media and representation in research* (2nd ed.). Sage.

Punwani, J. (2009, 10 October). The 70s rebels. *Times of India.* Retrieved from http://timesofindia.indiatimes.com/articleshow/5109980.cms?utm_source=contentofinterest&utm_medium=text&utm_campaign=cppst

Shaw, W. S., DeLyser, D., & Crang, M. (2015). Limited by imagination alone: Research methodologies in cultural geographies. *Cultural Geographies*, *22*(2), 211–15. https://doi.org/10.1177/1474474015572302

Sheller, M. (2014). The new mobilities paradigm for a live sociology. *Current Sociology*, *62*(6). https://doi.org/10.1177/0011392114533211

Thrift, N. (2000). Non-representational theory. *The Dictionary of Human Geography*, *88*(4).

3 Becoming a book lover

Mediating the present through the past

Buying books was the most obvious aspect of the *boipara*'s everyday routine. I have therefore always found it interesting that the regulars of the space are, nevertheless, always aware of the possibility of finding something unexpected or exciting from the plethora of second-hand and collected materials in the bookstalls. Choosing to be a regular visitor to the book shops *and* bookstalls of College Street required a certain seriousness, an earnest emotional investment. As a young girl, one of the activities to which I looked forward every year was the thrill of acquiring the *new* books for school. It was a feeling I shared with all my cousins growing up in our big, shared family home. The seeds of appreciating books were sown into our psyches from the outset. But when we were in the early grades of school, it was not just that they were books, nor that they were schoolbooks: they were *new* and symbolically stood for the beginning of a new grade. At the same time, I remember how I gradually started to look forward just as warmly to the old books. Although second-hand, they were still new to me. There was something special, quite different affectively, about acquiring these two generations of books. The new books came with a sense of responsibility and anticipation – I had to use them well to succeed in my schoolwork. The relationship with the second-hand books grew from a feeling of not knowing what Dadu would bring for me every year, a different excitement each time.

As well as the anticipation and excitement I felt at the arrival of these books – both new and old – there were also the feelings associated with having to give away the previous year's used books and exercise copies to the second-hand book shops. Over the years, this turned out to be a very personal, precious and treasured ritual between me and my Dadu. Recalling these annual rituals now, I think the circuits of exchange between us and the *boipara* involved more than the books themselves and that they expressed emotions and sensations that were not just mine. Dadu would take my old books, sell them to the bookstalls and bring the new books from the book shop for the new grade in school. My imagination followed the used books and exercise copies back to the *boipara* and would remain there with them for a time among the second-hand bookstalls.

DOI: 10.4324/9781003293026-3

Figure 3.1 Everyday regulars walking in between the bookstalls organised within the inner boundaries of footpaths

In the following years, I was struck by the very different ways in which the new books were bought from the shop by individual customers and the old books were bought by the bookstall owners from customers-as-sellers. In case of the new books from the book shops, the payment for the books and exercise copies had to be made to the shop based on the individual prices of each book. But the second-hand bookstall owners bought books *according to weight*, measured on old scales with the weights on one side and the books on the other. It was a very common practice among Calcutta's second-hand bookstall owners to weigh the total number of books, notes, magazines and so on and pay the individual sellers based on the going price of paper per kilogram.

Unlike the bookstall owners, Dadu was meticulous in his sorting and storing of reading materials such as the old newspapers, news magazines, literary and political little magazines, periodicals and books that he purchased and collected over time. I often wondered how much of all the storing and archiving he did was important; there were often complaints from other members of the family about how storage space at home was being used up by useless papers. However, my grandfather always had an explanation for this: every year he collected the daily newspapers, news magazines, little magazines, periodicals and books because he treasured what he read, and he believed that all the materials he collected should rightfully go to the second-hand book shops and little magazine stalls for others who looked for these materials to access. He often described the satisfaction he felt when he sold his collected materials to the *boipara* second-hand bookstalls. Even though, when I was still in school, I was never taken to the second-hand book market in College Street, there was a very strong relationship that was established through the travels of my school books, exercise copies and other materials every year. In many ways, my grandfather and I shared our emotions about books. Years later, during my university days, I would visit and sit[1] in the second-hand bookstalls that my Dadu used to frequent. I never expected that my books would still be found there, but there was a strong sense of remembering the space – remembering the *boipara* that I had never visited at a young age but had imagined vividly.

My visits to the *boipara* as a student, as a graduate and as part of this fieldwork often involved remembering the space: recollection of stories, anecdotes, events, moments – some of it had happened and some of it was imagination. I have realised that the act of remembering is not just bringing the memory of the past into the present. It involves a close knitting of the imagined and the actual, both experienced in the past. This idea of reworking is evident in the experiences and memories I describe above: my introduction to, understandings about and imaginings of the *boipara* have been through multiple trajectories. It was never one or another. As I remember, the affective resonances that I experienced during my first few visits as a college student were a combination of what I had anticipated the *boipara* to be like before I ever visited it and reimaginings of the multiple stories, anecdotes about it and lived experiences of it related by my grandfather and father. This contributed a sort of creation of impressions and events to my imagination, bringing into my imaginary life recollections of moments that I may not have actually lived in that past time but had regularly shared through *their* storytelling and *their* lived experiences.

There is also a very strong material connection – I have often reflected on why it was important for me to explore the bookstall to which my Dadu sold my second-hand books when I was at school. It was almost as if a part of my earlier self made the visits to those bookstalls and to the *boipara* in general. My later actual visits were thus mediated through a series of affective, mobile flows of memories, material associations and cultural, political and social associations I had developed with it through the stories I had been told and the histories I came across. That is, each time I experienced the space, I engaged in a process of reworking that added multiple layers to the fabric of the space through my actual presence, my affective imaginings, family history, collective memories, written histories and numerous other trajectories of sensorial, intellectual, creative and material experiences.

In interacting with the spatiality of the College Street *boipara*, I needed to consider whether we might develop a spatial aesthetics that allows imaginations to flourish, affects to resonate, objects to become things and memories to influence the making of the present. I was also looking for an ontological perspective on myself as local and researcher and visitor and writer that might enable me to reflect on my approaches to understanding this space while accounting for multiple desires and anxieties; allowing a sense of being that has no finality but is always taking place, sometimes drawing from imagined and lived memories, both of which are in dialogue with unfolding presents and immanent futures. A narrative understanding of space that is in motion, in process, 'always under construction' and always open to transformation through relationality cannot be contained in a fixed framework, structure, foundation or single all-encompassing idea, nor can it be based in any settled ontology.

When I returned to the *boipara* for my fieldwork, how its spatiality unfolded for me was not restricted to that time or to my experiences then. It seemed like a palimpsest of all the associations I had made with College Street: as a child and then as a teenager whose books were continually bought and sold there; as my grandfather's granddaughter emotionally and affectively linked to the circulation of his books, magazines and newspapers (and their part in the knowledge economy of the space); and as a young adult when I too came and went, actualising my longstanding imagined relationship with the *boipara*. Exploring the *boipara* for my research, I had to come to terms with this palimpsest and its many different layers, existing between the space now, me and the different senses of its spatiality that I had developed from my many previous encounters and the stories I had been told. In addition, I had to make a few other admissions to myself: now that I was studying the space, as well as experiencing it again, there was a sense of newness, as novelty, and immediacy to my experiences, but I also realised that my relationship with the space was at risk of becoming voyeuristic. Yet a somewhat voyeuristic role as an observer did not make me a mere spectator in my fieldwork process. I quickly noticed that everything I did while studying the *boipara* was also an act of anticipation and remembrance. This involved a sense of simultaneous rebuilding of the past and the present. There was, then, no alienation from the space involved in my position as researcher and no restriction on my present interaction with the space. I was perhaps more like a facilitator or a medium with whom multiple dialogues took place, and through whom many rhythms and unfolding(s) passed.

Donald (1997, p. 184) writes that, 'in order to imagine the unpresentable space, life and languages of the city, to make them liveable, we translate them into narratives'. He stresses that this production of narratives is not a mere accounting of images or imaginations but a 'projection of multiple selves: past, future, imaginable selves onto the cityscapes being recorded' (Donald 1999, p. 189). In this essay, Donald develops the idea of multiple narratives networking among themselves and the part played by certain stylistic elements of novel writing, like form, genre and so on in how cities unfold in narratives. He uses Virginia Woolf's *Mrs. Dalloway* and Alfred Doblin's *Berlin Alexanderplatz* to explore how multiple intersections between the past and the present, the imagined and the lived, create a narration that is *more than* representational. As he puts it 'the urban is mediated through a powerful set of political, sociological, and cultural associations' (Donald 1999, p. 179), which are brought to the fore as much through the expression as through the narrative *per se*. Donald's argument, then, is focused on how the expression and style of the narrative captures the imagined city. My trajectory was different. I became interested in encountering the *actual* surfaces and interfaces in the zone where multiple facets of the *boipara* meet each other as well as encountering something of the depths of imaginary and mythic relations, memories and stories out of which the *boipara* emerges. I needed to find a way to recognise and to write about the continual negotiations, coexistence and co-belonging of an array of emotional, affective, historical, imagined, remembered and lived continuities and disjunctions with the mobile and volatile present.

This was what excited me most about the possibilities opened out for me by scholars such as Jacobs, Donald, Highmore and Massey. If one cannot any longer claim to 'represent the real', how can a richly endowed space such as the *boipara* help one to write with, through and from a real that is always already mythic and imaginary; that is deeply politicised but refuses to be captured by 'history'; that one loves and in which one's own memories are implicated? And can the outcome present itself as 'research'?

As these aspects of my work settled into some kind of clarity for me, I became even more conscious of a sense of my multiplying possible selves: a flattened and linear representation of my experience(s) was impossible, but what was possible? I began to recognise that I needed to find ways of spatialising the *boipara* to facilitate different forms and systems of representation and conceptualisation. I wanted to unravel the complexities of the *boipara* without simplifying them. I needed to approach representational forms that uncovered the thick, mobile networks of interactions in the space. I also came to realise that the motivation behind writing this book was not really about representing the College Street second-hand book market; rather, it has always been about the possibility of experiencing that space and articulating it simultaneously.

Experiences of heritage never really have boundaries that define their identities. They are *only* identifiable by their continuous state of change or flux. Yes, there is repetition and routine embedded within everyday spatial practices, but every repetition adds something new to how the every day unfolds. It was important to question the privilege that comes with constructing the identity of a space. Thus, understanding of difference, especially in developing the concept of the assemblage through ideas of relationality and difference, becomes vital in how I seek to understand the spatial movements of the *boipara*, including those aspects of it that may seem to be distinguished by its repetitions. My exploration focuses on the intersections of the various interactions that occur *across time* between the material, the non-material, the affective, the emotional and much more.

The stories, experiences and anecdotes used throughout my book in unfolding my associations with College Street and the *boipara* do not necessarily fit themselves into any coherent trajectories of time. I set out to narrate experiential information in ways that enable the weaving of further connections.

In relation to my descriptions in Chapter 1 and earlier in this chapter, it is clear that when I came in contact with the second-hand books that my father and grandfather brought to me at a young age, my affective relationship with the *boipara* was being formed in real time while I was reading them. But this sort of experience could not be located in a simple, linear version of time. Given their encouragement and how their books circulated, I made contact with and began to understand a spatiality before I experienced it directly. In sharing with me small, even insignificant stories from their memories of College Street, Baba and Dadu were inviting me into their real-time experiences of the space in the past. As a young adult, when I became a regular customer of the precinct, my experiences were continually informed and shaped by multi-faceted successions of affects: those of my past imaginations of the space as a child, those of the stories I had heard of the space through Baba and Dadu and those I heard from other regulars once I became embedded in the space myself. When I later visited as a researcher, the layers across space-and-time deepened, thickened, and became richer and even more complex.

The stories I heard about the *boipara* created a sense of association and familiarity in my mind before I ever went there. They were also sometimes food for my imagination. When I did go there, I pondered these stories again, imagining in a newly informed way what could have been if I had been there at that time. My experience of time when I explored the space as a researcher was quite similar but not exactly the same. By then, my knowledge of the *boipara* and of College Street more generally was guided primarily by stories that I had encountered, sought out *and* created, and also informed by how I reflected on my earlier acquisition of stories and imaginings as well as how all of my early experiences – as a child and as a student – could slide into my current experience unexpectedly, *at any time*, as it were. The 'dimension' of the spatiality of the *boipara* had become and is still becoming a complex network produced out of what Massey (2005, p. 24) describes as 'multiple trajectories, a simultaneity of stories-so-far'. Thus, the moment we acknowledge that this spatiality is in fact by default a multiplicity of duration, we liberate the space from the sense of representation that comes with a linear and static approach to time. This multiple reworking and rebuilding of experiential knowledge and information seeks a form of expression that is itself liminal and in flux.

Notes on the margins

In this section, I examine the potential thinking of the spatiality of the *boipara* as an assemblage. Deleuze and Guattari (1987) develop the idea of the assemblage most fully in *A Thousand Plateaus: Capitalism and Schizophrenia*. The concept has been picked up in several disciplines and used in many ways by different scholars. While some of these appear as points of both convergence and departure later in the book, for the moment I am using 'assemblage' in a way that works best in understanding the *boipara*. To do this, I am engaging in a sort of thought experiment: combining several lines of thought that have been triggered by and through my experiences of the *boipara*. Rather than applying the notion of the assemblage as a ready-made concept, I need to trace the various conceptual threads that have helped me understand the idea, because that thinking and those threads are also part of my engagements with College Street. They are, in other words, already bound up in the multiplicity that is the convergence between the *boipara and* my experience *and* my understanding of it.

While browsing through second-hand books in one of the stalls, I noticed how some of the books have notes, personal messages and scribbles on them. I had seen these before on the second-hand books at home as well. These are different in every book. Some of them are personal notes, some of them are academic and some of them are illegible on account of having different handwriting. A far cry from the prevalence of sticky notes these days, I remember that taking notes on the margins, writing down a passing thought was common practice when we grew up and probably even before. I myself have these notes and thoughts scribbled all over my books and have left them on some of the second-hand books that I gave up at the boipara. *Here on my fieldwork, I have noticed plenty of notes. The thing about them is that they always make you wander into newer, unknown and different avenues, trajectories and trails of thought. However, I wondered what the booksellers thought about these. Rajen da, whose bookstall has a prime position on the footpath outside Calcutta University, provided some fascinating stories about these notes.*

Rajen da, a bookstall owner's comments on people's additions to books

Didi, did you know that some books, their stories and their extraordinary writings used to be very close companions to the regulars of the 1960s, 1970s and even the 1980s? After having read a second-hand book, someone would come and resale it to us. But the resale used to happen with a few conditions. They used to let us know exactly who would come and get the book next time. It was kind of an informal request from their side to sell it to the person they thought would come. Inevitably their partners or companions would come later in the evening and buy it back from us. These books would have personal often intimate, romantic notes on the first pages or sometimes even corners of some pages within the book folded for their partners to discover. But these were not the only kind. Sometimes groups of friends who believed in particular ideologies or just regulars of our stalls [who] were general avid readers of literature, politics, science or philosophy used to take great interest in this kind of circulation. One of the main reasons for this is that in boipara you never know what you chance upon. Thus, if anyone finds a book, diary or notes that they know is valuable to them or to their social circle, they make sure it is circulated. See, for the regulars of the boipara, the circulation is important, in fact more important than personal possession. You can say that this is a norm of the place. Here everyone shares books. Also, after everyone in the group had read the book, it used to be a topic of discussion or adda in the tea stalls and the coffee house.

Me: Wouldn't it just be easier if someone bought the book and gifted it to their loved ones straight away?

Rajen da: No but that is not the point. There is a certain sense of mystery, excitement and anticipation that is added to the process of rediscovering these books from the stalls again. For lovers and companions, it added a sense of mystery and complexity to their romantic communication. For other intellectuals, it was a sense of earnestness and allegiance to acquiring knowledge. We, the booksellers, often wondered about this and laughed amongst ourselves too. For the lovers, it was these sentences or group of words that they often dedicated to each other, by writing small notes and scribbles along the sides of these pages. Sometimes they would carefully fold the corner of these pages for the other person to know, sometimes they would have dry leaves as bookmarks inside them. Some people still do it now, but the practice is not popular anymore.

The ways in which the notes, scribbles and the stories on the pages of the second-hand books function is like the functioning of the *boipara* itself. I was used to the way the students left their study notes as marginalia and scholia. But what Rajen da described added yet another dimension that reinforced my sense of the similarities between the notes, the books and the *boipara*'s own spatial movements. Like the books themselves, their informal contents also become active participants in producing and carrying forward narratives, or suggestions of narratives in and about the space. The booksellers, the customers and the second-hand books participate in multiple rhizomic movements. Rajen da made clear in his stories about the notes and scribbles on the margins of the books' pages that, despite the fact that it might actually be practical for those people who were participating in the circulation simply to buy the books for the person they intended to receive them, they chose to use the multiple forms of expression, tones, languages and modes of connection that the books offered them in order to express themselves.

These processes are in some ways strikingly similar to how I experience associative thinking and imagining through the circulation between my experiential knowledge of the *boipara* and the many different ways in which I encountered it and reflected on those encounters as a researcher. I have described how, in visiting the *boipara* at different times, I carried forward bits and pieces of everything I had heard, read, thought and imagined about it to inform the experience I was going to have. When I reflect on these books, their writer's voices, the stories or scholarly discussions within them and the notes written about their contents or as a message (or both) by someone – from Rajen da's perspective, with relative frequency for a *specific* someone else with whom they were emotionally or ideologically connected – I wonder whether each of the books also informed individual subjectivities and enunciations. I also wonder whether the expressions, feelings, affordances and affects left by *boipara* regulars who choose to participate in this rhizomics of books indicate that they enjoy knowing that their voices are part of a multiplicity. Do they resist occupying an individual and singular position, becoming an element in a multiplicity while at the same time taking part in what are, in many ways, highly idiosyncratic and individuated shared practices? If, rather than just underlining or commenting on things they need to recall in relation to their own studies, they are participating in the forms of expression that Rajen da describes, then one imagines that they pick up a book they have read, find excerpts, paragraphs lines or even words that connect with them on a personal emotional and/or ideological level and use those parts or contents as a means of expression. This form of mobile expression is in itself complex and layered, but what interests me most is that all of these regulars, through their practices, made the *boipara* a part of a circulation of emotions, ideas and ideologies between people that extends far beyond the *boipara* yet simultaneously carries the *boipara* beyond itself.

So, after someone bought these second-hand books, read, connected and made their personal notes on them, they brought them back into the stalls to resell them to the booksellers. Most of the time, this was with the expectation that their loved ones, close friends or ideological comrades would come back to this same stall and buy back that same book. It is obvious that, between these people, and between them and the book sellers, there was some form of communication and mutual understanding that this would happen. Thus, the whole process takes on the aspect of a familiar procedure or routine. What, then, was the point of having this kind of ritualistic circulation with and within the *boipara*? On first hearing about these practices, I imagined that perhaps this mode of going about things had its roots in a time when it was a safer way to communicate, personally or politically – whether it was a relationship or an oppositional stance to certain regimes that made overt public contact or conversation risky. While it was hard to pinpoint the exact motivations, it was clear that the atmosphere, the environment and the occupants of the *boipara* are densely invested in a sense of keeping alive movement and active engagement, through sharing knowledge, reading and internalising it, and then engaging with it along with friends and companions collectively in the tea stalls, coffee houses and sometimes even inside the bookstalls and bookstores. For lovers, perhaps this circulation was shared more privately, especially if it facilitated a relationship that was otherwise not socially and/or culturally acceptable.

There are a few complex modalities of interaction at play here. First, I note that by communicating one's personal feelings, emotions and views through the second-hand books, the regulars are indulging in a complex 'weaving in of narratives and voices'. Second, they are voluntarily engaging in a mode of communication that blurs the lines between the real and the imagined. Third, throughout the whole movement of this communication, the booksellers, the books and their contents, the participants (i.e. the friends, comrades or lovers) tread through a path that is characterised by a sense of in-betweenness, of knowing and not knowing, of being known and not being known. All this adds to the richness of their relationship with the *boipara*. They purposefully make

Figure 3.2 The bookstalls well lit with bulbs and tube lights in the evenings

it a point to involve the books, the stories and information in-between them. For them, the beauty, the excitement, the usefulness, the value was *not* to achieve clarity or singularity in their expression; it was to actively engage and express themselves through relatively chaotic, complicated, deliberately messy trajectories of circulation.

For Deleuze and Guattari, rhizomes and assemblages are co-constitutive and assemblages are in turn co-constitutive. In relation to living things, Claire Colebrook (2002, p. xx) explains assemblage thus:

> All life is a process of connection and interaction. Anybody or thing is the outcome of a process of connection. A human body is an assemblage of genetic material, ideas, powers of acting and relation to other bodies… There is no finality, end or order that would govern assemblage as a whole; the law of any assemblage is created from its connections.

The *boipara* is characterised by temporary and ongoing encounters and coexistences of many, many narratives and materialities and subjectivities and bodies and experiences. The movements involved in rhizomes and assemblages frequently operate across multiple strata in multiple directions with multiple continuities and discontinuities. So why does any assemblage form? As Deleuze and Guattari (1987, p. 399) address this question of process, it is a matter of passions and desires:

> Assemblages are passional, they are compositions of desire. Desire has nothing to do with a natural or spontaneous determination; there is no desire but assembling, assembled, desire. The rationality, the efficiency of an assemblage does not exist without the passions the assemblage brings into play, with the desires that constitute it as much as it constitutes them.

Part of the compositional motivation for the *boipara* stems from the strong sense of intellectual engagement and affection we have noticed between the regulars and the *boipara*: the processes that the regular book buyers use involving various kinds of attachments – as demonstrated by my Dadu, for example – with second-hand books, their contents and their toing and froing from the *boipara*; the commitment of the booksellers themselves to their occupation, their regular customers, their books and the *boipara*'s place in the life of Calcutta; and in some instances, as Rajev da and others discussed, relations with the circulation of book that are actuated through a strong sense of passion in the form of emotional, political or ideological commitments. However, there is also a strong desire connected to being there, propelling a sense of movement that continues to bring the material parts and the participants of the *boipara* together. The connections that each of these parts and participants make are rhizomic. Through the course of each of these movements, their purposeful and random points of specificity meet, creating affective connections.

A rhizome comes from the *milieu*, referring to the middle, the in-between. The middle is thus the generative space from which other movements take off, perhaps causing ruptures, perhaps coming back around (a rhizome always returns to itself), perhaps making new connections. This generative ferment means that the overall movements of a space do not cease or fall silent, and it is very difficult indeed to suppress or silence them. Rather, there is continual production of significant points of departure and connection so that new connections are always immanent. Assemblage is therefore always taking place, shifting, reconnecting, about to take place in different ways on different trajectories and so on. For example, when the contents of the second-hand books become active participants in the kinds of communication processes discussed above, the lines or words that are dedicated from one person to another become modes of enunciation *and* potential points of connection. In exploring notions of collective assemblage, Deleuze and Guattari (1987) stress that when it comes to language as expression, to matters of signification and subjectification, it is not the individual but the social that matters. Noting that 'relatively few linguists have analyzed the necessarily social character of enunciation', they continue:

> The problem is that it is not enough to establish that enunciation has this social character, since it could be extrinsic; therefore too much or too

> little is said about it. The social character of enunciation is intrinsically founded only if one succeeds in demonstrating how enunciating in itself implies collective assemblages. It then becomes clear that the statement is individuated, and enunciation subjectified, only to the extent that an impersonal collective assemblage requires it and determines it to be so.
>
> (Deleuze & Guattari, 1987, pp. 79–80)

In the case of the circulation of communications via a person's purchase, resale to the same stallholder, a different person's purchase, a different person's inscription or mark and resale in the same stall with the tacit understanding of the bookseller as to the next purchaser and so on, there is a 'collective assemblage' that pre-exists any particular exchange of this kind and at the same time emerges from every exchange of this kind. However, it can hardly be considered to be 'impersonal', informed as it is by understandings between various people who undertake different roles in the circulation. It is, nevertheless, indisputably *social*. Such movements of enunciative acts via the books are themselves different elements of a collective assemblage in particular parts of the social (e.g. couples, political groups, friendship groups), while the *boipara as a space* is also a collective assemblage of which the many different social arrangements producing the circulation of modes of enunciation are in turn a part, as are the stalls, the books, the street, the movements of books and people and many kinds of knowledge in many forms to and from College Street. All these communications, exchanges, tacit and formal encounters – material, embodied, intellectual, political and affective – *make multiplicities*. That is, just as multiplicities make connections, so too do connections make multiplicities. This obvious characteristic of complex spaces highlights the massive degree of over-simplification necessarily involved in attempts to reduce the complexity for the sake of analysis using particular representational strategies that take the form of 'tracings' in Deleuze and Guattari's (1987) sense.

But to return to the fascinating forms of communication that Rajen da explained to me, when the person for whom a particular second-hand book is destined picks it up and reads the lines, there is a dynamic point of connection which might at once be thought of as a both pause and a rupture, during which new noticings can take place, and things which are immanent can become real. This one moment in the middle of the complexity of the *boipara* is grounded in desire, passion, intellectual, political and/or emotional investment, depending on the reason/s for the communicative exchange. The potential rhizomic connections implied by the enunciation or expression in the specific pages, sentences or words have been made particularly meaningful and at the same time they impel further rhizomic connections. They become potential, even probable enablers of departure. In due time, a predictable destination for those books, the notes in their margins, the underlined phrases, is that they will become part of someone else's lived reality. If those enunciations do acquire an individuated, subjectified meaning – insofar as the 'impersonal collective assemblage' at that time 'requires it and determines it to be so' (Deleuze & Guattari, 1987, p. 80), it will occur because the books and the underlining and inscriptions in them enable lines of flights. While continuously changing its own nature, any assemblage also enables such lines of departure that have the effect of continually creating newer stories, narratives and affective responses which 'diffuse' from the *milieu* in the moment of rupture (Deleuze & Guattari, 1987).

Just as assemblages are products of and producers of connections within and beyond themselves, so the assemblage as a thinking tool becomes useful when it enables the researcher to notice how certain movements involved in pauses and ruptures, connections and disconnections open out potentials for further actual and imaginative encounters and engagements.

Street names and associative narratives

As well as the major buildings and stalls in the *boipara*, the College Street area also has a park with a public swimming pool, in an area called College Square. The name 'College Square' is also usually extended to take in the streets and lanes on the eastern and western sides of the park. This group of streets has a number of significant buildings, including the University Institute, the Mahabodhi Society Hall, the Theosophical Society building and the Students' Hall. These additional buildings of the education quarter act as the border delineating the College Street precinct as a whole.

It is an interesting, yet very common, characteristic of Calcutta that, over time, many streets, lanes and by-lanes have found themselves with multiple names: sometimes they retain their colonial hangover by holding on to the English names. The street on the southern side of the College Square park is called Mirzapur Street, although as one moves west it is known as Kolutola, while further along to the west the name changes again to Canning Street. People have their own preferences, too, regarding how they name portions of this street. It is impossible to pin down a single reason as to why different sections of the street have different names. The reasons are varied and multiple because the affective associations that people have with these spaces are also varied and multiple.

During my fieldwork, I became involved in a casual conversation with a bystander in this area. At the time I was conducting an interview with a stationery shop owner about the emergence of stationery shops and their coexistence with established bookstalls. The owner observed that this had occurred as a gradual progression rather than a sudden proliferation. He also pointed out that since students need stationery, the growth in stationery shops was inevitable. The stationery shops, though, are spread along the *Kolutola* area, the part of the street most often known by its Bengali name. Our conversation turned to the multiple names for parts of the same street and the implications of this. In another example of the unpredictability of what can become 'data', it was then that the bystander joined in.

Comments from a bystander

Didi, I have been overhearing your conversation for a while. I have been a regular of this space for a long time as I work in one of the paper binding factories nearby. A part of my job is to come to these stationery shops to supply notebooks and exercise copies to these shops every other day. I have noticed that very rarely do people associate these names of the same street because of the literal purpose of a street name. There are so many ways to remember a street name other than that. Like I know some people will never call Kolutola Canning Street. It is because of a language sentiment. If you ask why, they point out Kolutola being a Bengali name makes way more sense to them than Canning Street. It's Bengali, not only in its literal existence but also in its historical, cultural essence. It gives a homely feeling to the street. I have heard endless people stand at the corner of the footpath of this street and argue in an evening adda *session over cha and cigarette, as to why it makes more sense to call the street Kolutola rather than Canning Street. On the other hand, there are so many others who would rather call the same space Canning Street. The British have left us long ago, but some people love to hold on to the memory of the same. They wonder how the name came into existence: who may have named it, was there ever a Lord Canning? [laughs]: just about anything to get them going with the* adda. *However, my point is, Didi, I don't think you can find a single reason behind this idea of multiple names for the same street.*

Here we can see that the coexistence of multiple names for the same street demonstrates the social character of the *boipara*, made readable through a collective assemblage of enunciation.[2] To break this down, let us consider the ontological fabric of the *boipara*. There are a number of layers of 'being the *boipara*' that are at work here. In terms of the physical structure of the space, as the bookstall owner observes, there is always a proliferation, extension, gathering, alignment and growth of the contours of the *boipara* by virtue of everyday usage and the realities of consumption. The boundaries of the space keep changing. Some like to acknowledge the growth and sprawl as a crucial part of the space. Some do not. There is no clearly demarcated boundary of this space; it is 'generative' and it is subject to relational understanding (MacFarlane, 2011, p. 650). Thus, on the physical level, exactly which streets, lanes or by-lanes constitute the *boipara* is relational, socially fluid, subjective and also idiosyncratically and ideologically *personal* to everyone who uses this space. Each of these streets has numerous historical, political and cultural stories attached to it. They have come to mean and/or be emblematic of different ideas for different people.

For some, the street of the stationery shops is always going to be Kolutola and for others it will be Canning Street. For these people, their sense of place-making will depend on their own collection of personal stories. There is no sense of conflict or duality in the existence of two (and possibly more) street names. During my fieldwork, I found myself in a kind of nomenclature 'in-between' in which I did not have to choose between the two names. I was simply being made aware of the existence of multiple, small everyday understandings of one of many parts of the *boipara*. Because I had not previously given this much thought, this knowledge has now spilled into my own previous understanding of the *boipara* and invited me to ask further questions. For example, after I was made aware of the existence of multiple names for the stationery street, I started wondering, 'How many other such sites of multiple names exist in the *boipara*?'.' I kept thinking of the many new and different ways in which my spatial engagement with the *boipara* could activate and unravel the existence of my personal accounts of placemaking and those of others. In his article 'The City as Assemblage: Dwelling and Urban Space, exploring the potential of the assemblage as a conceptual tool through which to think about city space, Colin MacFarlane (2011, pp. 649–67) writes:

> Assemblage orientates the researcher to the multiple practices through which urbanism is achieved as a play of the actual and the possible, and as such resonates with the broader histories of critical theory and critical urbanism …
>
> Its object is therefore the emergent, the processual, and the material, the city that is being made through the active and disparate labour and resource that aligns in particular ways and that is constantly subject to being imagined and lived differently.

The *boipara* as assemblage, or several inter-implicated assemblages, continually creates these movements of real conditions of possibilities, lines of flight, moments that impel the creation of newer narratives. The sensations, affects, ideas or imaginings produced by such interactions are not limited to their immediate interpretation: their power lies in their openness to being perceived in innumerable ways. This, in turn, makes possible a plethora of newer narratives that can immediately destabilise the existing assemblage and open it to further connections.

Being open to deploying the concept of assemblage as a thinking tool necessitates becoming comfortable with realising that one is *setting out* to produce multiple ways of thinking about spaces as well as the circumstances, events and impacts of their becoming-otherwise. Simultaneously, the same openness ensures that one *notices* how assemblages make multiplicities by becoming open to more stories, anecdotes, experiences, events, interactions and connections. To notice is to become alert to what is immanent as well as what is redundant. In the *boipara*, production of multiplicities takes place through the booksellers, buyers, everyday regulars and visitors; the material, affective and sensorial components; and the distinct languages, idioms and other modes of enunciation that characterise the space. As a researcher who understands the potential of my work to be disruptive-and-productive, my own thinking, conversing and writing processes become implicated in how the assemblage might (and might not) change; how the circulation of pluralities of many kinds connect (and do not connect) with each other. In drawing attention to these many components of the *boipara* and to how one can work with it/them as an assemblage, my purpose is not simply to present/represent (and thus 'demonstrate') the complexity and richness of what is there and what happens there; it is *also* – importantly – to indicate the extent to which the movements of people, including me as insider, participant and researcher, and people's memories, experiences, myths, personal and cultural narratives, meetings and communications, become part of the happening – part of the eventness of the space. Everything constituting and participating in the *boipara* as assemblage impels it – habitually, routinely – to reiterate its movements, events and intensities while also creating ruptures or disjunctures, either occasionally or frequently, depending on social and political circumstances across time. So everything and everybody is implicated in assembling, reassembling and producing new assemblages within or beyond the assemblage/s that one has noticed.

Thus, with regard to Rajen da's account of various kinds of 'clandestine' book circulation in which booksellers were/are complicit, the question that enables us to understand the spatial movements of the *boipara* as assemblage is not 'What do these second-hand books or their contents represent?', but rather 'What do these second-hand books *do*?' This opens other questions such as 'How do they activate the space?' and 'How does this impact the way the assemblage forms new connections; transforms; produces further assemblages?' Similarly, the multiple names for streets and lanes in the *boipara* and neighbouring precincts, and the multiple strands of stories, histories and myths variously used to 'explain' those names, do not simply provide insight into various perspectives about history and culture, thus providing insights into how historical events and socio-cultural networks intersect in urban space; the names and their many subjective associations activate those everyday lanes and by-lanes in ways that make them *performative* of the space, its stories, its movements and its connections within and beyond itself.

It seems to me that we can understand the active circulation that *boipara* regulars participated in by making the contents of the second-hand books part of their lived experience in the space and beyond it as an example of 'practical action' that enables me to make observations about it in relation to 'theoretical action'. There is a relay produced and new networks created between practice and theory as well as between practice and affect, practice and politics, practice and ideology, which become part of how the *boipara* understands and mythologises – even theorises – itself across time and space. The capacity of the contents of the second-hand books to interact with other components of the space imparts a sense of agility and mobility to the overall communication process. Thinking through assemblages thus 'allows us to attend to how these often-disparate activities become entangled with one another, but nonetheless have potential agency beyond those interactions which may later become parts in other assemblages' (Anderson et al. 2012, p. 173). When we think of going to the *boipara*, the first activity that comes to mind is either buying or selling books or other materials. We do not necessarily take into account the act of getting to the *boipara* or the anticipation that builds up during the movement towards the space. Then, once we reach the *boipara*, we start to attune ourselves to its existing interactions, rhythms and movements. We interact not only with the books but also the booksellers; we think of sitting in the bookstall or spending time in the coffee house. On the surface, all these decisions, arrangements and rearrangements of expectations and plans may seem 'disparate', but they are all entangled, informing our senses, our movements, and hence our reactions in the space.

In his translator's introduction to *A Thousand Plateaus*, Massumi (1987, p. xiii) observes: 'A concept is a brick. It can be used to build the courthouse of reason. Or it can be thrown through the window.'

Tools build, repair and extend things, and also deconstruct, undo and demolish things. The concept of assemblage, like any tool, is thus an active means to assist us in building, dismantling and rebuilding things, structures, situations and ideas. Tools only exist because of their uses: we invent them for their capacities. Assemblages as thinking tools enable newer connections and modes of thinking while also changing themselves. Sites, spaces, places, people, activities, routines, material things, institutions, grand buildings and ramshackle structures, histories, memories and stories are all there in the College Street precinct. However, they are not in fact *there* as an assemblage, or a collection of interconnected assemblages, until *we notice* their potential to be there as such and to be productive and disruptive once apprehended in that way. Unlike tools whose basic configuration changes little across time-and-space due to the specificities of their uses – knives, stirrups, the ladder, the screwdriver, the hammer – assemblages are conceptual tools that inevitably change and change themselves in the moment of becoming-tools, of making connections – that is, of generating consequences.

Assemblages are fragmentary because they are produced by what we notice as present-and-immanent. They are always in transition, characterised by the connections they can make with an exterior continuously transforming also due to being characterised by fragments, fissures and ruptures. Such processes of relationality are why assemblages become apparent and are always becoming-multiple. Thus, 'the assemblage' cannot be treated as a pre-existing framework that would fit the *boipara* in order to delineate its contents, its boundaries and how we can represent it. Rather, the concept of assemblage is relevant to the *boipara* because we recognise it as a valuable tool that can help us to work with and write about the complex nature of such a space. We could say that, in this book, 'assemblage' acts as a conceptual consequence that enables other conceptual strands to emerge from my consideration of the space, its movements and its significance for the city in which it is situated.

Notes

1 The bookstalls, although makeshift and shy of space, often have small wooden stools, benches and other moveable temporary sitting arrangements inside and around them. These benches, tools and so on are often even placed on the footpath close to the stalls.

2 In Chapter 6, I explore how certain words and parts of compound words that characterise the language of the space, like the *boi* and the *para*, facilitate 'incorporeal transformations' within the space. In doing so, I explore how the *boipara* functions as a collective assemblage of enunciation.

References

Anderson, B., Kearnes, M., McFarlane, C., & Swanton, D. (2012). Materialism and the politics of assemblage. *Dialogues in Human Geography*, *2*(2), 212–15. https://doi.org/10.1177/2043820612449298

Colebrook, C. (2002). *Gilles Deleuze*. Routledge.

Deleuze, G., & Guattari, F. (1987). *A thousand plateaus: Capitalism and schizophrenia*. University of Minnesota Press.

Donald, J. (1997). This, here, now: Imagining the modern city. In S. Westwood & J. Williams (Eds.), *Imagining cities: Scripts, signs, memory*. Routledge.

Donald, J. (1999). Imagining the Modern City, Minneapolis: U. 1999.

MacFarlane, C. (2011). The city as assemblage: Dwelling and urban space. *Environment and Planning D: Society and Space*, *29*(4), 649–71. https://doi.org/10.1068/d4710

Massey, D. B. (2005). For space. *For Space*, *1*, 1–232.

Massumi, B. (1987). Realer than real. *Copyright no. 1*, *2*, 90–97.

4 Noticing the background-moving through affective environment

On walking with the background as a research background

As with any space, describing the *boipara* is not an easy task. What do I notice when I look at it as I move around the area? What decisions do I need to make about what my sense of the space includes and what remains outside? And what do I place in the foreground or in the background, emphasising or limiting its importance? The *boipara* is not composed of isolated or exclusive collections of objects or people; what composes the *boipara* is often beyond the perimeter of the physical or semiotic limits of the objects and subjects only. The parts and participants of the *boipara* are always in movement both in terms of where they may be found physically and what they mean at any given point of time. They are in continual conversation with the aural, sonic and material actors of the space. Thus, the mood or atmosphere of what the *boipara* is, moment by moment, always spilling into/becoming-something-other than what it seems to be. If this is the case, then material objects and human subjects cannot sensibly be picked out and described as the individual components of the space through which we can make sense of the whole. Within the *boipara*, what matters is not what its parts represent, but what they are capable of doing, producing and transforming into while they are always becoming-something-else.

This chapter concentrates on affect, exploring sensorial experiences and their expression through smells, tastes, colours, sounds and touch. I develop a more sensitive practice in relation to our noticings, feelings and senses and carry these into our thinking and writing practices of the *boipara*. The chapter examines the nature of walking in the market as a political act, and it also explores and reflects on the potential of walking as a practice-based research methodology. Tying in the experience of walking and wandering in conjunction with the sensorial registers of smell, colour and soundscape within the *boipara*, the chapter demonstrates how the second-hand book market works as an affective and sensorial assemblage, giving the space its unique rhythm and cultural fabric.

The *boipara* is marked by circulation, change, to-ing and fro-ing of bodies and books. People come to the *boipara* daily to buy second-hand books and bring other books for reselling. Those who buy books take them back to their

DOI: 10.4324/9781003293026-4

homes and make them part of their own lives. They use the books in diverse ways in their personal contexts. They read them, scribble on them, make notes in the margins, put in chits of paper as bookmarks and as notes to themselves for study or remembrance, and sometimes – much to the dismay of the booksellers – they tear pages from them. After various periods, they will probably return books to the *boipara* for further use and circulation. Similarly, the contents of the bookstalls change regularly. As the existing second-hand supply goes out, batches of other second-hand books, with their notes and other supplements, come in. They are rarely placed in the same order in the stalls: they are put in different piles, on different racks and are produced for customers in new contexts and for new and different needs. In their unpacking each morning and packing away each night, the books daily experience 'mobile-spatial encounters' (Barry 2017, p. 33). Similarly, the stalls themselves change. The yellow and blue paint with which many of their tin walls are painted needs to be renewed now and then. Some stalls are whitewashed. Some booksellers like to have wooden stools and chairs in their stalls, while some simply sit on a pile of books. The booksellers often have other random materials and objects – or,

Figure 4.1 Piles of old and new books waiting to be displayed in the stalls

as they call them, tools that help to change the boundaries of their stalls. For example, sometimes they will use bamboo sticks and plastic sheets to extend a stall's perimeter. This, they told me, usually happens at those times of the year when the buying and selling frequency is higher than usual – at the beginning and end of the academic year, or a new term, for instance. Sometimes they make these changes because of the weather.

The stalls are not only used to store and display the books, but they are also a place where regulars can spend time – an arrangement that meets a sentiment and understanding shared by booksellers and buyers. Thus, the booksellers are continually organising and rearranging the interiors of their stalls to make them more usable. Similarly, the customers do not merely come, collect their books and leave. They spend time browsing, chatting with the stall owner and other customers, drinking tea and clearly creating an experience for themselves. For the booksellers, how they experience the customers experiencing their stalls changes with each change they make to the structural layout, the positioning of a stool or bench and the arrangements of books.

Taken together, the bookstalls, the neighbouring eateries, the shanty tea stalls, the outer edges of the walls of the universities and other buildings, the railings along the footpaths seem to express a shared spatial understanding. For example, the tea stall owner never complains about a customer who always sits and reads the daily newspaper or books from a neighbouring stall in his meagre space. Similarly, it is completely against the culture of the coffee house to ask someone to leave on account of their having spent too much time occupying a table. There are no formal rules regarding how the space should be used, but there are always deep-rooted tacit agreements, affective understandings, emotional investments and conscious actions in play about how the spatiality of the *boipara* is created and recreated. It is a vernacular understanding and production of spatiality grounded in practice, repetition and difference that creates a unique 'atmospheric' environment.

It is about one o'clock in the afternoon and I am walking along the footpaths of the boipara. *Walking and the scorching heat of Calcutta summer never go hand in hand, but as far as the* boipara *is concerned, this is an exception. Everyone who visits the* boipara *loves walking here. In my previous walks in the space, I have experienced this too. However, today I feel overwhelmed and spoiled for choice in terms of the trajectory of walking I should begin with. By virtue of my multiple experiences of walking in the* boipara *at different points of time, I now know of several ways of walking in the space. One of these trajectories was more exciting than the others. Apart from coming to this space as a beginning regular with my college and university friends, I did make a few visits to this space with Dadu, right after I finished school. Dadu always navigated his way into the* boipara *through the* daftaripara.

The *daftaripara* is 'the backyard' of Calcutta's publishing industry, known for its concentration of binding works, printing presses and other print-related small-scale enterprises. The cluster of lanes and by-lanes that form this section of the city often overlaps with the physical area of the *boipara*. One end of the *daftaripara* shares a border with College Street and the other end is near the Sealdah train station, which connects the city with the other outer suburbs. I think, for Dadu, getting me introduced to the *daftaripara* simultaneously with the *boipara* was very important because he always had a penchant for knowing and being a part of what goes into the making of books. I remember, walking along those printing presses and book binding presses, how Dadu's face used to light up when he heard the huge letterpress machines printing 'thick retro style fonts' for both books and other commercial purposes like calendars and paper boxes for sweet shops.

Dadu's relationship with the loud noise of the letter press machines, the tiring labour of book binding and his close friendship with Iqbal, Ahmed and Rajen Kaka, who were a part of this old school printing industry was heavily emotional and is almost haunting for me today. Later on, when I would sometimes take his trajectories by myself, I would often wonder about the intimate relationship Dadu shared with these factories and their workers. Still undecided about how I want to walk along the boipara *today, I wonder, are Iqbal, Ahmed and Rajen Kaka still there? Do they still catch the train from the outer suburbs of the city daily and come to work in those factories? I remember, later when we would finally reach the* boipara *from the* daftaripara, *Dadu would point out, 'These books that you see, the notebooks you use, all get printed, hard bound and prepared for us to use in those factories on the* daftaripara.*' I cannot help but reflect on his reverence towards those printing machines, the thick ink and the black-bordered hard-bound covers. I wonder today what kind of emotional relationship he had not just with the books, but with the process of printing words on them, the quality and texture of the paper, the ink, a sense of reverence towards those huge printing machines. The second-hand books were important to him, but so were the printing press with all its material components, his brief stopover and five-minute chat with the owners and workers of these presses and, most importantly, his repeated practice of telling and re-telling the mechanics and history of how books and notebooks come into existence as we made our way into the* boipara *through the* daftaripara.

Still walking around the boipara, *I make a note to myself: maybe I could walk back home from the* boipara *through the* daftariapara. *This makes me both excited and uncomfortable. Is it going to be the same? If I reverse the trajectory, will I still find the same people, at least catch a glimpse of them working? Are those huge letterpress and printing machines still there? What if it is all different? A strange sadness almost makes my heart sink. But what if it is exactly the same after all these years? I am excited again. I will never know if I never walk.*

Our engagements with our spatial surroundings are always fluid, contextual and relational. As Massey (2005, p. 8) points out, 'We develop ways of incorporating a spatiality into our ways of being in the world.' When we do this, we do not pick and choose some parts of the space and leave out the others. My own experience with the *boipara* is similar. As I discuss elsewhere, I started to *make sense* of the *boipara* through a combination of the multiple stories, the experiences of the people I met and the people I knew, as well as my own knowledge, imaginings and speculations. When I think of the *boipara*, I do not just think of the booksellers and the second-hand books. My senses also open up to the colonial buildings, the interiors of my university, the routine to-and-fro movement of the trams, the sounds of the students chatting while leaning on the railings of the outer edges of the footpath and the numerous other incidents, episodes and interactions that take place there. Conventionally, these elements would simply be considered background, as if in a film or photograph. However, in the *boipara* everything seems to command attention. The human, material, affective and sensorial interactions within the space create a unique effect where any boundaries between the background and foreground blur. It is as if differing focal depths disappear in terms of spatial relations and in terms of mood and atmosphere.

There has been a growing interest in the idea of atmosphere within a number of disciplines (e.g. see Adey, 2017; Anderson, 2009; Anderson & James, 2015; Ash, 2013; Stewart, 2007; Bissell, 2010). As Anderson and James (2015, p. 34) note, '[T]his literature emerges from, and seeks to develop, existing work on affect and affect theory – itself an increasingly prominent series of theoretical trajectories for analysing how life is organized outside of strictly representational registers and structures of meaning. If we think about how spaces can organise and express themselves beyond the conventional terms of representation, we can no longer take for granted that the background is merely a passive scene against which interesting features of a space become noticeable. Thus, we need to think of what role the 'background' plays in what takes place and in the production of affect. In this connection, Seyfert (2012, p. 29) asks two pertinent questions:

> How can an affect be simultaneously defined as an effect that only emerges from the encounter between bodies, and *also* as a force external to these bodies? In other words, where does affect *begin*?

While I cannot provide comprehensive answers to these questions here, I can point towards a line of inquiry worth considering. In thinking about the background of the *boipara*, I wonder how the background aids in experiencing the heritage value of the foregrounded character of the space – the books and the book market. Making *sense* is momentary, affective. It is like there is a pause in the constant processes of change and movement that distinguish the *boipara* (or any other space) during which things simply *make sense*. The affective experience begins the moment we start to make sense of our animated atmospheric environment. The challenge that this poses is how to notice and recognise the background and its part in the creation of atmosphere or mood in a space. Anderson and James (2015, p. 35) note, '[T]he background has been given a number of names such as milieu or context.' While these terms acknowledge something about the participation of background in a scene, their potential has not been explored fully. For most people, background remains largely passive, even as 'context'; it remains the setting against which representational descriptions of objects or animals or people become possible. However, I think that keeping this challenge of the background in mind allows us to work towards experiencing and understanding the particular spatial atmosphere to the *whole* of the *boipara*. The moment of making sense contributes to the realisation of this specific, affective atmosphere or mood to the space. It is impossible to pinpoint at any moment when or how that atmosphere is generated, since it is grounded in continual creation and characterised by repetition, mobility, difference and relationality. But for me, what is important is understanding that this atmosphere marks the interaction and influence of the background with the foreground in the emergence of a sense of the spatiality of the *boipara*.

Walking hand-in-hand, with the background

Walking is an embodied activity of the observer. It is difficult to think of it as something taking place in the background. There are numerous ways in which walking insists on being noticed. For example, in a previous chapter I touched briefly upon how walking towards the *boipara* would inevitably 'interfere' with my fieldwork plans for the day. I began to realise, however, that walking was not external to – and thus a distraction from – the processes of observation and interaction with the precinct, but in fact made an essential contribution to these. In effect, my steps can be said to have functioned in much the same way that background does (a similarity underlined by recognising and realising how 'unnoticed' walking becomes the more familiar we are with an area). Thus, after a few days' thinking about how to make sense of my walking itself (and its interferences), it occurred to me that it was, in fact, *part* of my sense-making process. When I think of walking in the *boipara*, it is a composite and complex act. Yes, it takes place as movement along trajectories, both forwards and backwards, but it also involves pausing, loitering in bookstalls and stopping altogether. Moreover, when walking through a space, we are always adapting ourselves to it, attuning ourselves to its rhythms, synchronising ourselves with the movements of others. It is a matter of moving *with* the environment, not simply in it. Peter Adey's (2006) idea of 'relational mobility' is useful here. Adey observes that it is equally important to think of 'immobility' and 'fixity' as spatial conditions if we want to grasp what movement entails. For Adey, the relationship between immobility and mobility emerges as follows:

> In short, the politics of mobility revolve around two main ideas. First, that movement is differentiated – that there is a politics to these differentials. In other words, that power is enacted in very different ways. And second, that it is related in different ways, it means different things, to different people, in differing social circumstances. In other words, mobility and immobility are profoundly relational and experiential.
>
> (Adey, 2006, p. 83)

With this perspective in mind, two points need to be made about the act of walking in relation to my discussion. The first, which I have already started to frame, is that in the context of the *boipara*, walking is neither a peripheral nor a background activity. Second, along with it being relational and experimental, it is also composite and interactional. A few walking episodes may help to reflect on these points.

Sujan is a second-year economics student at Scottish Church College. Scottish Church College is not located in the boipara *area. It is a 20-minute walk away from the precinct. However, Sujan is a regular of this space as he visits the area at least two to three times a week.*

Sujan's observations

Sujan notes how even though it is faster and convenient to hop onto the tram or a bus to get to the boipara, *he prefers to walk. He reckons it is a lead-up to being in the* boipara. *He adds that his knowledge of the space has expanded from just walking. He had heard about a few bookstalls and stores from his friends during his initial days of getting to know the* boipara. *At that time, it was only a matter of going to the* boipara. *Gradually, he developed an interest on 'getting to know the place'. Walking helped with this. He discovered numerous lanes, by-lanes, eateries and general hangout places other than the coffee house, from just walking.*

Another diary entry records my conversation with a student from Presidency College, who drew some interesting connections between walking and belonging.

Koel's observations

When I first came to this college, like many of my friends, I had to figure out a way to become a regular of the College Street, you know. There is a sense of belonging that my seniors and others felt that I struggled to feel in the beginning. So, I made it a point to walk in and around the boipara *every now and then. It was similar to putting myself into the practice of walking. Then, one fine day I got it! I could see and experience why the* boipara *was so special and important to my friends.*

Something happened between Koel's beginning to walk in the *boipara* and her realisation that she was part of the space. Her own walking allowed her to notice *how* others used the space. This helped her develop her own pace, and thus her own relationship to the area. In addition, she noted how she would not have known about certain lanes and by-lanes without repeated walking and exploring the *boipara*. Koel also told me how, over time and through her own rhythms – walking, stopping, measuring her own pace and directions – she was able to find her own set of favourite bookstalls. She observed:

> *I was never bored of this practice of walking because I wanted desperately to be a part of this place. I did not just want to come and study in Presidency University, I wanted to be accepted by the secret norms and practices of the* boipara. *I wanted to have my own stories.*

Unlike her friends, she did not hang out much in the coffee house. She preferred another smaller eatery that she found while walking. Thus, her sense of place-making and belonging was directly related to her own understanding of the rhythms, atmosphere and pace of the area. But these experiences were not all uniquely hers. Some moments in her sense-making were collective, being shared with friends. This fostered her sense of belonging. But there were other moments when a developing sense of feeling comfortable in, and at one with, the *boipara* was hers alone and intimately connected to the trajectory of her own walking. I want to suggest that this kind of belonging through place-making is grounded in the experience of what I have called *making sense*. This is momentary and fleeting, but it is also productive in that to apprehend *a* sense in the multiple movements, weavings and networks of an environment or place allows us to recognise the possibility of other engagements and other moments of making sense. In other words, any recognition of belonging that arises in a moment of sense-making looks towards future engagements and what they might mean for us – and how the space is making sense for us just as we are making sense of it. We can begin to imagine or conjecture about what might happen next (something that needs to be distinguished from prediction since prediction tries to acknowledge and comply with the limits imposed by reality, the laws of probability and calculation). In the context I am considering, imagination and conjecture may be tied to lived reality; however, they are also open to possibility and to creativity.

The idea of pacing is also important in thinking about the *boipara*. Its environmental conditions situate and tacitly organise us by facilitating as well as restricting our movements. As Mackenzie (2002, p. 122) writes, '[W]e have no experience of speed except as a difference of speed.' Similarly, we have no sense of trajectories without differences in trajectories produced by spatial arrangements, people, objects, obstacles and attractions. Interacting with the booksellers, the books, noticing the architectural arrangements and the general movement in the *boipara*, we continuously negotiate with the 'mood' of the space while reorientating our speed of walking. Consequently, walking is not a straightforward, constant mobile activity but, as I have observed, includes pausing, noticing, adjusting and readjusting speed and pace. Several times, I have walked along the *boipara*, but each time it has been different. There is always something new happening to make me pause and take notice and think – like a new group of students protesting on something specific to *that* time, or a new bookstall that may have been set up, or sometimes even a rearrangement of a known bookstall or eatery or hangout place. I would not have noticed these changes without walking, of course, but I would not have been able to *make sense* of them without stopping and recognising something different. In this sense, walking is like weaving with movement and time, with materiality and human responses and human interaction.

Walking in the *boipara* can thus be described as composite and interactional. While, in a literal sense, walking refers to movement, in terms of the interactions it makes possible, it is also and always a negotiation between mobility and everything that immobility can disclose. If walking also entails our responses – as the continuous orienting and refiguring of the relations between thought and environment – it is not something to be regarded as merely peripheral to experiencing the *boipara*. It is, in fact, central to *making sense* of our embodied experiences of the site. David Macauley (2009, p. 21) writes:

> The French poet Baudelaire also suggests ways in which an urban walker might deliberately abandon his or her guard so as to be seduced, moved, excited, saddened and eventually abandoned by a chance encounter or fantasy with a beautiful stranger or intriguing passer-by on a busy sidewalk.

While I have been stressing a connection between noticing objects and events and qualities (and thus a kind of awareness) and making sense, it is important to recognise the significance of letting one's guard down in how noticing takes place. Both Sujan's and Koel's reflections suggest how their place-making through walking begins to involve the letting down of their guard as their attention to the practice of walking is displaced by familiarity with the *boipara*. De Certeau (2002) regards walking in the city as a kind of writing and rewriting, a meaning-making process that allows us to orient and reorient ourselves in the space of the city. But while, as with writing, de Certeau makes walking a more or less semiotic activity, making sense is perhaps broader in the kinds of meanings it allows us to consider. Movements (like walking), materiality and human and non-human interactions are continually creating 'relational networks of meaning and belonging, of time and space' (Duff, 2010, p. 890). As far as walking in the *boipara* is concerned, the kinds of meanings that emerge through making sense involve how we understand our affective responses to the environment itself and to its objects and arrangements, and how we respond to relations with others: the booksellers, students, regulars, visitors. That is, it is a matter of how we become familiar with the place while remaining open to the many new and different experiences it continues to present.

Soundscape and colour

Sound, colour and light play an important role in the mobile, relational and shifting atmosphere of the *boipara*. Whenever I begin trying to describe how the spatiality unfolds, multiple visual and auditory elements come to mind. These are not confined to books, bookstalls, conversations I have had or friends I have met there. I also recall the colours of the paint on the stalls, the fading yellow paper of the musty-smelling second-hand books, the smog from the traffic on the main road cutting across the precinct and the clunky clatter of cars and vans on the cobbled surface, the sound of the tram, the calls of the hawkers and stallholders offering bargains, the clamour of protest marches – or *michil* in Bengali, the school and college students chattering and laughing, the clanking of tea urns and tin mugs from the tea stalls, the very different noises of the coffee house with its echoing, spacious interior and high ceilings, the cross-currents of many conversations round wooden tables, teaspoons in ceramic cups, the waiters' voices, their shoes on the wooden floor, and the extraordinary impact of their nineteenth-century uniforms on the overall ambience of the place. These things all stay with me. The outer walls of the university, college and school buildings are covered with layers of posters and street art, whose combinations of colours, words and parts of words and images remain to be recovered as palimpsestic fragments by a mixture of memory and imagination. For anyone who has been there regularly, wandered about and lingered in the space, the word *boipara* is enough in itself to evoke a complex array of sensory impressions.

Observational and reflective diary notes

I notice the rhythms of the space attuning to my own bodily registers the moment I cross the footpath from Bidhan Sarani main road into the boipara. *These days, almost all the bookstalls keep at least some new books along with the old ones. However, I remember Dadu mentioning how there was a time when the stores had only second-hand books. Those of us who are current and/or the have-been regulars of the* boipara *are very aware of a distinction between those who buy new books and those who buy old books. It is kind of understood in the space that those who buy new books – that is, apart from one's curriculum requirements – were always looked down upon by those who only bought old books. I had been a part of this trope for a long time, lecturing my juniors in college how they were doing a disservice to the space by buying or looking for new books that they need, and not the second-hand books that they really should be looking for. This was an important stand that we thought we were taking for the* boipara. *However, apart from our anti-capitalism stand of supporting bookstalls over bookstores, there was something else that drew us emotionally closer to the bookstalls. The stalls provided us with a familiar environment. We knew the sound of the voices of the bookstall owners, the familiarity of the chatter of* adda *of our friends who used to sit in the stalls, discussing life for hours had a sense of knowingness that was comforting. We always had our own* thek, *an informal Bengali expression the owning a space for a few hours by a group of friends. These* theks *(spaces of casual hangout) were usually the bookstalls or the tea-stalls.*

As young adult, one of the things I was consistently curious about was that the boipara *was always so special to the regulars, like my Dadu, not only for the circulation of second-hand books, but for the circulation of knowledge – through the books they read, the intimate familiarity they formed with the sounds and colours of the space and the conversations they had there. It was strongly evident in the vivid descriptions they gave of the space. Over the years, I got to be a part of this circulation as well.*

I remember, as college students, my friends and I started wearing khadi *clothes and using rug bags, especially if we were visiting the* boipara *area. Some Bengali novels documented how the youth of the 1970s were especially inclined to wear clothes made of locally produced cotton and* khadi *material: attached to this dress code was the idea of a middle-class Bengali, reformed, experiencing a cultural renaissance and politically left-leaning. Times had, of course, changed by the 2000s when I went to college. However, even today, as I sit and write notes about the* boipara*, sipping coffee in the coffee house, I am wearing a chrome yellow* kurti *with jeans. I cannot help but reflect on my choices then (that is, during my college days) and even now. There is something more than a socio-political and cultural statement happening here. I am looking around and I notice that there is a strange composition or a gradience of colours within the coffee house. It is hard to describe yet vividly experienced: the rugged brown wooden tables and chairs, the deep brown coffee, the partially slipping-in sunlight through the skylight windows on top of the coffee house roof, the grey, dense layer of smoke that looms and circulates in the room. In our daily lives, we come across innumerable material objects. Among these, different spaces always have specific ways of making some of these material objects 'their own'. I remember Dadu always using ink pens and HB pencils for writing or taking notes on the margins of his books. In the coffee house today, I notice that there are no laptops or tablets, but notebooks, pens and pencils that people are using in the coffee house now, as am I. I can see that a lot of the students are carrying their laptops but not using them. I notice that the dark blue and green colour of the ink from the ink pens complements the chrome yellow and brown hue of the room beautifully. I had never noticed this before, but today on my fieldwork I see how all these colours coming together gives me a sense of familiarity and belonging.*

It is obvious that my experiences are attended by strong associations with the sounds and colours of College Street. It is not only the sensory impressions that matter but also the affects they produce, the embodied responses, the emotions. For example, many of my interviewees commented how the coffee house creates an environment of comfort and a feeling of 'being at a place that we have known for long' through the sound of multiple conversations, groups of friends singing together with a guitar and the coffee table tops as their musical accompaniments, while the coffee continued to flow. Homagni, now pursuing a medical career, was a student at the medical college in College Street in the 1990s. Her recollections of that time are vivid, and she made a point of mentioning how she remembered the space through colour and smell.

Interview comments from Homagni

Boipara r amej tai alada *[The atmosphere of the* boipara *is something else]. The peculiar smell that used to come from in between the old pages of the books, the chrome yellow colour of the pages were half the reason why I used to buy second-hand books. There is also something really grey and gloomy about the* boipara. *I cannot really figure out why exactly this colour comes to my mind right now. Is it the streets or the smog? I remember the broken footpaths kind of having the same colour as the cobbled tram lines. The bright yellow cabs, the white and blue tramlines, and the uneven-shaped stalls added a unique texture to the place. Then there is the sound of the electric spark from the antenna and the connection lines of the tram. The trams are significantly slow compared with the surrounding traffic. You know, I remember the buses, taxis and the private cars moving through the traffic at a completely different pace than the tram. The tram is slow and halts every five minutes in the College Street area, so I remember the trams more vividly. Also, the protests and vigils that took place near the area added to the traffic. It was so chaotic and slow and congested, but I loved it. Somehow everything seemed to work. The congestion, the composition of colour, the sound and the overall interaction created a unique mood.*

Like many of my recollections with which I commenced this section, Homagni's description is focused on qualities and impressions: on shape and movement and colour and sounds rather than descriptions of objects and people (e.g. it is the trams' slowness and their relationship to the relative speed of other vehicles that she remarks on in recollection). Her descriptions begin to evoke the atmosphere of the place and how it envelops her memories ('*It was so chaotic and slow and congested, but I loved it*') and dwell on moving, seeing and noticing the *mood*.

Experience in a space and the elements that comprise the sensorial atmosphere share a symbiotic relationship. I have already explained that, despite the habit's of representational expression suggesting otherwise, in fact elements like sound or colour are never simply 'in the background'. They have active and tangible effects on how we navigate and how we register what is occurring in a place. For example, while I remember many occasions when I walked along the lanes and by-lanes of the *boipara*, often stopping to browse books in the stalls or have a conversation if I bumped into someone I knew, there were also many other times when I would simply stop and gaze at the area and all the movements taking place around me. Sometimes these pauses were in response to events: the sound of chants and slogans being yelled from vigils or marches somewhere unseen, which prompted me to ask someone passing by whether they knew what the protest was about; sometimes for merely practical reasons on reaching the edge of the road where I would need to wait for a break in the chaotic traffic. Sometimes I seemed to stop for no particular reason. Whether one's movements and pauses are casual or deliberate, the momentary 'eventful' unfoldings of a passage through the *boipara* are neither empty nor passive. Whatever interrupts us or distracts us opens our engagements with colours, sounds, smells, speeds and intensities that influence our own movements, responses and understandings. That is, when we stop we do not stop only to notice, but to *think*. These are the very moments when colours, sounds, smells, speeds, intensities and contrasts become active in influencing our own process of movement.

At the same time, these also become the moments when we stop to *make sense* of what may have invited us to stop, when we reflect on the books or the stalls or the traffic or the coffee house. In doing so, we begin to realise how the atmosphere of the place communicates with us, and to respond to that. This is strongly akin to the recognition in exploring the relationship between atmospheric movements and tourism that certain destinations 'intensify our attunement to the collaborative processes of re-orientation during transit' (Barry, 2016, p. 375). Even without 'a' destination – indeed, because of a multiplicity of destinations – this is true of a visit to the *boipara*. What differs, of course, is the degree to which, in the *boipara*, this collaboration is interpreted by those engaged in it as an outcome and affirmation of their long-term embeddedness in the space. The affective collaborative relationship is powerfully associated with being 'a regular', in recollection and in current experience.

Tim Edensor (2015) uses the idea of the city at night as a 'second city' to explore the logic of oppositions between light and dark. The term 'second city' comes from Sharpe's (2008) description of a place '"with its own geography and its own set of citizens" that emerges when daylight fades' (Edensor, 2015, p. 423). Edensor finds in this idea the basis for a critique of conventional cultural constructions of darkness. He also notes how lived navigation through, and experiences of, dimly lit spaces can produce affective experiences that are far removed from these conventional understandings (Edensor, 2015, p. 423).

In relation to books, the question of whether a space is well or poorly lit clearly becomes important. There is normally an expectation that any place involved in the reading and buying and selling of books will be adequately lit. This is a matter of both physical convenience and health. A dimly or poorly lit library or book shop is criticised for ignoring these concerns. In the *boipara*, the more permanent and established book shops, and the publishing houses, have always been brightly and uniformly lit. The dimly lit bookstalls, although they have now become better lit due to changes in street lighting technologies, and increasingly have even more bright interior lighting using newer, smarter modes, are not designed to facilitate reading. Their cluttered, cave-like arrangements more often than not still render reading difficult – the piles of books can themselves block their new lighting from various angles, and that lighting can in any event be less effective at different times of day, in different weather, in different seasons. In the end, though, all of that is a non-issue for most patrons. The numerous, old-fashioned and/or inefficient lighting set-ups that still pertain in many stalls, and the newer forms of lighting in other stalls that are ineffectively placed in relation to readerly needs, continue to provide an atmosphere for the stalls, and thus and for the *boipara* generally, which regulars tend to treasure. This is because dim lighting contributes to a sense of intimacy, excitement, serendipity and anticipation that people associate with the *boipara*, a place they have long known or have more recently come to want to know precisely because of the atmosphere and mood associated with it.

Inside the bookstalls, there is little room to move. Unless situated at the end of the footpath, each stall is hemmed in on both sides by neighbouring stalls. Their inner walls are lined with shelves for displaying the books, while the space on the footpath immediately in front of each stall is covered in piles of books. The clutter of books and magazines, and the cramped nature of the stalls, reinforce the impression of how poorly lit they are – even in the daytime.

Figure 4.2 Stall-like structures emerging out of book shops

This is certainly something that was obvious to the stall owner I cited in the previous chapter, whose recollections of the *boipara* were framed by his experience of the history of public and stallholder lighting there, and the impacts of that on the bookstalls. His point in rehearsing that history, though, was that people seemed to prefer dim lighting.

Drawing on the work of Bryan Palmer (2000), Edensor (2015, p. 429) notes that darkness or the absence of light can be interpreted as a time for

> transgression, as well as disconsolation and alienation, the "time for daylight's dispossessed – the deviant, the dissident, the different"… the opportunity for clandestine, revolutionary and conspiratorial activities that foster the imaginative, creative "resources of otherness" that challenge the daytime norms of commerce, economic rationality and regulation.

I have already discussed in relation to assemblage, in particular, how the *boipara* is a space that continually opens out to other spaces, producing movement, flow and flux. One of the examples I used of the connections beyond itself that the *boipara* makes were the insights provided by bookseller Rajen da regarding the fascinating range of 'clandestine' uses to which the circulation of second-hand books were put in the past, from secret love affairs to underground political activities. We have also seen that the *boipara* is still sought out and 'protected' by regulars for its bohemian ambience, its lengthy associations with all kinds of cultural, social and political resistance and rebellion. Further, even if it is a place where 'the daytime norms of commerce' apply, there is a kind of irony at work in the extent to which many aspects of the space actively promote those 'night-time' activities of imagination and creativity.

While there are apparently ironic disjunctions between the cheek-by-jowl mass of cluttered bookstalls that constitute the crowded pavements of the *boipara*, and the intellectual, social and cultural fluidity and *openness* of the College Street area to difference, creativity and political radicalism, this would be a mistaken reading. For those who enjoy the precinct, the co-location, connections and circulations involved in the College Street assemblage(s) produce coherences and ruptures *at one and the same time*. They notice cracks that serve as enablers of departures, encouraging lines of flight, imagination, exploration. These, in turn, circulate into the city as a whole with the people and books moving between. I suggest that the students, scholars and other everyday regulars of the space who continue to visit these bookstalls feel that an over-abundance of light creates a disruption to their familiar experience and that it is that very familiarity that makes the *boipara* feel welcoming. Light, or the inadequacy of it, is simply another element to reinforce the ongoing sense of difference, resistance, counter-cultural thinking and unconventional social attitudes sought by people in and from the *boipara*, and that the space itself contributes to producing, materially, affectively. Taken together, the contrasts, coherences and ruptures characteristic of the *boipara* are what people make use of in making sense of it, and in finding, by doing so, that it offers them opportunities for choice, expression, creative action.

The inclusive disjunctions (Deleuze & Guattari, 1987) of the College Street precinct are crucial to understanding the relational nature of space as well as looking at 'the things that happen in space' as events. They are equally crucial to understanding the *boipara*. The experience of browsing at any time through books in the confined physical space of a bookstall, surrounded by a blue and grey tin wall, for example, and the musty smell of second-hand books, peering through the low light at their yellowing pages, undoes conventional oppositions between light and dark, blurring the differences between night and day for the customer. These closed-in spaces in fact become sites of flux and of difference, venues for the movement of imagination, for openings. It is no surprise, then, to discover that the regular customers who still come to these stalls are disturbed by the encroachment of too much light. There is more than nostalgia to it. Their sentiments point towards another level of experience of the *boipara*, measured by affect in the contradictory ('transgressive') circulation of sensory perceptions.

Coffee house and *adda*

Like the stalls and their layout, the books and magazines, the movements and colours and sounds of the space, the *adda* forms an important aspect of the *boipara*. Historian Dipesh Chakraborty (1999, p. 110) explains, 'The word *adda* (pronounced "uddah") is translated by the Bengali linguist Sunitikumar Chattopadhyay as "a place" for "careless talk with boon companions" or "the chats of intimate friends".' *Adda* is a unique practice essentially involving friends coming together in various public and private spaces as a part of their everyday routines. The origins of the *adda* are difficult to pinpoint but, as a practice, it typically associated with Bengali culture. Chakraborty (1999) writes vividly about the practice of the *bangalir adda* (the Bengali *adda*), observing that at the time he was writing at the very end of the twentieth century – and this is still the case – the practice of the *adda* is taken to be 'peculiarly Bengali and … marks a primary national characteristic of the Bengali people to such a degree that the "Bengali character" could not be thought without it' (Chakraborty, 1999, p. 113). As Chakraborty observes, contemporary Bengalis can be identified as Indian (living in the state of West Bengal) or Bangladeshi citizens. My references to the practice of the *adda* reflect only on what occurs in India (indeed, in Calcutta). Indeed, my attention is focused on the *adda* not so much as an example of a Bengali practice but more specifically on how it manifests in the *boipara*.

In his preface to *Kolkata'r adda*, Samarendra Das (2010) writes that there is a specific sense of judgement or sociological baggage associated with the word *addabaaj* (one who participates in and is 'addicted' to the *adda*). It is as if anyone dedicated to their work should never fall into the practice of the *adda*. The reasoning behind such a judgement is grounded in the very nature of the practice. The *adda* is, however, a productive social practice through which many Bengalis have raised and apparently solved many social, political and cultural issues. It is where groups of friends come together to share their own stories, comment on other stories, enjoy the conversation and the company of others, debate issues and relish the process. The *adda* traditionally belongs to a number of spaces, both public and private, and it cuts across all classes and social divides. On a regular Calcutta afternoon or evening, one can witness *adda* unfolding in the *ro'aks* (which I described earlier); in the living rooms of Bengali households, at the corners of lanes and footpaths, inside shanty tea stalls and, most popularly, in the coffee house in College Street *boipara*. The regulars of the *adda* in the coffee house take special pride in their activity. It is as if the coffee house is their own place. Also, to participate in the *adda* in the coffee house has certain cultural, political and intellectual prestige (due to the location of the 'Albert Hall' coffee house in proximity to the most prestigious universities and colleges in the city).

I have explained that the *boipara* has long been the breeding ground of leftist progressive political ideologies, music, art cinema, literature and visual arts. As the current manager of the coffee house, Ratan da, proudly stressed in talking with me, the *adda* sessions of the coffee house have produced some of the most celebrated Indian literary figures, including Sunil Gangopadhyay, Shakti Chattopadhyay, Shankho Ghosh and Tarapada Roy. Gangopadhyay, along with friends and fellow poets and novelists, Ananda Bagchi and Dipak Mazumdar, was the driving force behind establishment of the celebrated literary magazine *Krittibas*. Commencing publication in 1953, it was the first of its kind to come out of the many informal *adda* that took place in the coffee house. Later literary movements that raised their voices against the establishments of their time, such as 'The Hungry Generation', also found and developed their initial voices through interactions in the *boipara* (as well as other settings).

Ronojit Das (2010, pp. 173–76) notes that every Saturday the huge hall of the coffee house, during the 1970s and 1980s, used to be packed with regulars who came for the *adda*. This was the time when Das was active within the literary scene and was also a regular of the coffee house. At that time, when Shakti Chattopadhyay was the young rising star of anti-establishment poetry, he used to make an appearance every now and then. His unbelievable grasp of and talent with the Bengali language made him the centre of all *adda*. Generally speaking, however, the coffee house was never a 'star-struck' place: when I was a student there, it was still common for poets, filmmakers and other artists to make frequent visits without that causing any fuss among other regulars. Also, nothing was off the table for discussion in the coffee house: from Marx to Kafka; from the films of Godard and Truffaut to Eisenstein's *Film Sense*; from local politics to crucial national issues; intractable social problems and the urgent need for social and cultural change.

Most of the coffee house regulars were chain smokers and the ashtrays on the tables overflowed with butts and ash. Over time, the place developed a permanent, distinctive odour consisting largely of a combination of cigarette smoke and coffee. The smoke is probably why everyone remembers the atmosphere as grey and foggy. I remember, too, how the loud discussions around various tables would cut through the dense air. While in the daytime the skylight windows created a greyish-white hue within the room, in the evening the atmosphere felt more intense from the yellow bulb lights. The lighting was uneven, however, being much brighter in the corners of the room but dimmer, with a more 'bohemian' effect, in other places. In effect, it was a theatre of a very different kind.

Das (2013) writes about how the coffee house regulars had preferred dress codes. Versions of these were adopted by the older middle-class intelligentsia and by young middle-class students as becoming-radicals. Cheap half-sleeve shirts, old trousers, rag shoulder bags and slippers were a standard dress code from the 1970s, and touches of it continued, as I have mentioned, to my own time as a student. Das notes how in the 1970s he had a weakness for jeans and thus his staple outfit was torn t-shirts and jeans, a look that he connects with his efforts to write sexually adventurous poetry. Many a time in the coffee house, he was criticised both for wearing jeans and for relying on sexuality in his poetry. At that time, despite the pride it took in its radical intellectual and political reputation, Bengal was still suffocated by a bourgeois discomfort regarding sexual expression in creative work. Das was one of the first poets to break this barrier. In his poetic career, he was both applauded and criticised for this.

Nevertheless, if you were published in one of the prestigious little magazines at that time, you were *always* applauded. The Saturday *adda* sessions in the coffee house were also important for this reason. Writers could bring in copies of their newly published work either in books or little magazines and distribute them to everyone sitting on the different tables. It was an effective way of getting one's name more widely known at the time.

Apart from the coffee house, the inside of some of the bookstalls, the publishing houses, tea stalls and even the corner of the lanes and by-lanes often served as sites for *adda*. These *adda* were slightly different to what took place in the *adda* of the coffee house. The *adda* of the streets, footpaths, stalls and lanes were not all intellectual and literary. They were more varied, with topics ranging from football and cricket to politics and general everyday observations of the weather or the traffic. People who had their own groups and specific spaces in the *boipara* made it a point to get together in these spaces and then, in due time, disperse. The *adda* were never peripheral to the 'main' activities of the *boipara*, nor were they simply background or an interruption to those activities. The prominence of the coffee house *adda*, in particular, and their place in the intellectual traditions of the *boipara*, of Calcutta, and even more widely of India meant that those unceasing, noisy conversations developed their own movement and connections beyond the tables where they happened. But, whatever their effect in the wider world, the everyday fabric of the *boipara* is unimaginable and incomplete without the *adda*, both in the coffee house and in many other places throughout the area. One can think of or be a part of multiple sessions of *adda* in the city, which means that the *adda* has always existed, regardless of its close relationship with the *boipara*.

Each of the elements of the *boipara* discussed here has scope for a much more detailed exploration. However, even if they are only fragments of a much more detailed account, it is these bits and pieces of the coming together of the apparently useless magazines, the sounds, colour, smells, the act of walking and the *adda* that all participate vitally in the spatial atmospherics of the *boipara*. Our experiences within any spatial environment is always composite when we remember a particular moment or period of time we spent somewhere. Thus, for me, writing about the *boipara* entailed witnessing or immersing myself in the shifting and complex theatricality, performativity, eventness, materiality and affectivity of the space. It is only in that way that I am able to find ways to understand the simultaneity and intricacy with which space undergoes continual construction.

Thus, it seems to me more useful to revisit, remember and rethink a place and its spatial movements than try to enumerate, analyse, demystify or deconstruct its complex networks of interaction. Sensations and movements are always congealing with each other while informing, reinforming and reorienting the space and its components. In an environment like the *boipara*, there is little to be gained by sorting all its elements into intellectual piles and ordering them according to arrangements that rely on a foreground and a background or any other hierarchy. As human, material bodies, when we do not resist the collective movements of a place and allow ourselves to be open to the unfolding of its sensations, affect starts to make sense and we start to make sense of it.

References

Adey, P. (2006). If mobility is everything then it is nothing: Towards a relational politics of (im)mobilities. *Mobilities*, *1*(1), 75–94. https://doi.org/10.1080/17450100500489080

Adey, P. (2017). *Mobility*. Routledge.

Anderson, B. (2009). Affective atmospheres. *Emotion, Space and Society*, *2*(2), 77–81. https://doi.org/10.1016/j.emospa.2009.08.005

Anderson, B., & James, A. (2015). Atmospheric methods. In P. Vannini (Ed.), *Non-representational methodologies: Re-envisioning research*. Routledge.

Ash, J. (2013). Rethinking affective atmospheres: Technology, perturbation and space times of the non-human. *Geoforum*, *49*, 20–8. https://doi.org/10.1016/j.geoforum.2013.05.006

Barry, K. (2016). Transiting with the environment: An exploration of tourist re-orientations as collaborative practice. *Journal of Consumer Culture*, *16*(2), 374–92. https://doi.org/10.1177/1469540516635406

Barry, K. (2017). Mobile-spatial encounters. In *Everyday practices of tourism mobilities: Packing a bag*. Routledge.

Bissell, D. (2010). Passenger mobilities: Affective atmospheres and the sociality of public transport. *Environment and Planning D: Society and Space*, *28*(2), 270–89. https://doi.org/10.1068/d3909

Chakraborty, D. (1999). *Adda*, Calcutta: Dwelling in modernity. *Public Culture*, *11*(1), 109–45. https://doi.org/10.1215/08992363-11-1-109

Das, S. (2010). *Kolkata r adda*/Gnagchil.

Das, S. (2013). The politics of agitation: Calcutta 1912–1947. In S. Chaudhuri (Ed.), *Calcutta: The living city vol. II: The present and the future*. Oxford University Press.

de Certeau, M. (2002 [1984]). *The practice of everyday life*. Trans. Steven Rendall. University of California Press.

Deleuze, G., & Guattari, F. (1987). *A thousand plateaus: Capitalism and schizophrenia*. University of Minnesota Press.

Duff, C. (2010). On the role of affect and practice in the production of place. *Environment and Planning D: Society and Space*, *28*(5), 881–95. https://doi.org/10.1068/d16209

Edensor, T. (2015). The gloomy city: Rethinking the relationship between light and dark. *Urban Studies*, *52*(3), 422–38. https://doi.org/10.1177/0042098013504009

Macauley, D. (2009) Walking the city: An essay on peripatetic practices and politics. *Capitalism Nature Socialism*, *11*(4), 3–43. https://doi.org/10.1080/10455750009358938

Palmer, B. (2000). *Cultures of darkness: Night travels in the histories of transgression (from medieval to modern)*. New York University Press.

Mackenzie, A. (2002). *Transductions: Bodies and machines at speed*. Continuum.

Massey, D. (2005). *For space*. Sage.

Seyfert, R. (2012). Beyond personal feelings and collective emotions: Toward a theory of social affect. *Theory, Culture, Society*, *29*(6), 27–46. https://doi.org/10.1177/0263276412438591

Sharpe, W. C. (2008). *New York nocturne: The city after dark in literature, painting and photography, 1850–1950*. Princeton University Press.

Stewart, K. (2007). *Ordinary affects*. Duke University Press.

5 Materiality and the *boipara*

Affective resonances and materiality

Writing about the *boipara*, I notice some curious movements, oscillations and transportations in my imaginative engagements with the space. They never take place in one or two directions, or pause on one or two preoccupations. They take place in multiple directions and intersect with formal and informal buildings, and with multiple aspects of infrastructure and materiality. They are neither stable nor static, which is to be expected – memory, after all, is a malleable and constantly changing process, and so is the real. Each of them shapes and reshapes the other. During my writing, when I think back, reflect upon, reconsider and re-evaluate my experiences of the *boipara*, I readily find myself navigating through the streets, lanes and by-lanes, leafing through the books, sometimes pausing to sit on the wooden stools inside the bookstalls, breathing in the smoggy air, listening to the familiar chatter, registering the sounds from the cobbled road. The tram passes every 15 minutes, cutting through the heart of College Street. Now and then, groups of students and scholars enter and leave the universities; however, none of them leaves the precinct. They always linger in the area, sometimes heading to the coffee house or the tea stall, sometimes hovering around the bookstalls engaging in casual conversations. I feel as if I vividly recall these everyday movements. Yet, while I am sure that some of my memories recapture exact experiences, I am aware that others *seem* real although they are almost certainly imagined and seamlessly interspersed with the experiential memories.

This chapter considers how the material and the non-material interact in the ways in which everyday spatial practices of the *boipara* reveal themselves. The chapter explores how the material elements of the assemblage function as more than cultural artefacts or products readily open to the subject's interpretive capacities. Focusing on the possibilities of knowing and doing heritage in the context of the second-hand book market, it explores the roles of affect in transforming objects into things within the assemblage of the *boipara* as a heritage site. The chapter begins by carrying forward discussions around affective resonances and materialities. It then envisages the space of the *boipara* by tracing the material connections made by the users with the objects and things in

DOI: 10.4324/9781003293026-5

the space. The final section explores the idea of the *para*, a Bengali word for neighbourhood, and its material, sensorial and cultural power to preserve and circulate the everyday practices and narratives of the neighbourhood of books.

Once I began to visit College Street *as fieldwork* – that is, *for the purposes of* research – I might have become even more worried about the possibility – indeed the likelihood – that I would 'miss' something important. This might have made my movements more purposeful and my observations more attentive. Indeed, for a very brief time it did. In my actual encounters in and with the field, this soon changed, and it continued to do so. Throughout my research from then on, and especially since I have been writing this book, I have reconsidered and repositioned that mixture of immediacy and anxiety, replacing it with recognition of the benefits of constant, flexible movement *and* of conjecture, of happenstance and imagination. I have understood the productive value of casting around in what has been, what is and what is immanent to what is – that is, what is becoming-otherwise.

The unsettling of the 'real' and of our connections to place by the apparently untrustworthy interventions of memory and imagination raise a further question: when we identify as being a part of a space – for example, a regular or, like me, a regular-becoming-researcher – with what do we most associate? What is preserved through our past experience through our material and non-material interactions that resonates through the present and anticipates future encounters in that space? And does that which is preserved add to our sense of *belonging to* the space? When we belong to a space, our experiences are enabled by and express our familiarity with it. At the same time, we acquire a set of skills – such as ways of being in and writing about the space – that help us to connect other narratives to that space and carry our narratives of that space elsewhere, into other engagements. It is a storyteller's licence stamped *by the space itself*. We can see this in operation in the following encounter.

From my diary notes

Today I talked with a second-year undergraduate student of Presidency University. Our interview venue was a combination of three sectors in the boipara *precinct. We began our conversation inside the canteen of Presidency University, and after about 15 minutes, Sohini suggested that we walk and chat. With me following her, we left the university campus and walked along the footpaths. After a while, we sat again inside a tea stall perched on the end of one of the footpaths and continued the conversation. However, the exchange quoted below took place within the university campus. I initiated our interview by asking her what it meant to be a part of the* boipara *and a student of Presidency University.*

Sohini: *Well, all my school life I carried this huge pressure to get into this university. There was this constant scrutiny on my grades by my family members; I had to always do well enough to ensure that I got into Presidency University.*

Me: *So what is this fascination with getting into this university?*

Sohini: *My family's obsession with me getting into this university was so closely associated with a sense of belonging and intellectual evolution. For example, my mother, who is an ex-student of this university, described how I needed to walk along the corridors of this university, spend time in the canteen, in the library and the bookstalls, be an active part of the students' union to understand politics, society and life in general. There was a clear need to belong to the materiality of the university and the* boipara, *to be a regular at the coffee house sessions. A few days ago, there was a police charge brought against one of the professors of the political science department. He was being monitored by the university administration for the longest time. When things got a bit out of hand, other professors strongly condemned this monitoring of teachers and their political ideologies. As a part of the protest, our professors conducted classes outside the classrooms for a week. We had our classes in the university playgrounds and lawn. These areas became emblematic of free, liberated education. Taking class in open-air sectors of the university gave a strong message to the administration. The university belongs to us, our minds and political sense of self develops here. We should have the right to choose what we align ourselves to politically. However, I specially connected with how we became creative in asserting our sense of belonging: by using the campus space outside of the classroom as a sign of protest. Simply boycotting and stopping class would mean we were giving away our right to the space. We did not do that: we asserted our sense of belonging by using alternative areas within the campus.*

What this first part of our conversation raised for me was Sohini's strong sense of 'ownership' of and of belonging to her university and to the precinct in which it is located, as if the two are inseparable in any 'real' sense. It was the same for me when I was a student at Presidency.

In this context, we need to think of belonging not as a 'state' or an 'identity', but as a *process* that manifests itself as both 'sited and mobile' (Lorimer, 2005, p. 87). There is no denying that the idea of belonging has been tied tightly in cultural, political and sociological analysis to questions of identity and textuality. However, belonging within a space like the *boipara* appears to be an opening to something more than that which is already there. The ways in which I understand and continue to experience the space are grounded in the fact that I once *belonged to the place*, physically and imaginatively. Later, during my fieldwork, I encountered several stories, interpretations and observations like Sohini's through which other regulars of the precinct made sense of their belonging through descriptions of processes that for them exemplified the nature of that belonging in relation to politics, history, spatiality, social and cultural forces and so on. In the course of my writing, I came to understand that the processual nature of belonging often opens out as well to creative thinking and storying around the past and the present of this space. However, before addressing my own processes of belonging in more depth, I want to discuss further the various interactions between the material, human and more-than-human components that contribute to a sense of the *boipara*'s spatiality.

The material speaks: Assemblage and writing about place

The interactions and emotional flows within the *boipara* take place not just between humans passing through – regulars of the market, booksellers and tea stallholders; they also involve its material components. Over time, the musty yellow pages of the second-hand books, the streetlights, the cobbled street, the tram lines, the coffee house, the bookstalls, the wooden stools and benches have become 'regulars' of the space as well because they are so strongly associated with it in the impressions of it that people carry elsewhere. The networks of information flow in which people participate are often characterised and produced through continuing interactions with the material objects of the *boipara* in producing its spatial assemblage. This chapter focuses particularly on interactions between and relationships with the material and non-human aspects of the space. I consider how some of the material components of the space and the multiplicity of narratives they produce contribute to the understanding of the *boipara* as an assemblage that is always becoming an *enabler of departures*.

The flexibility in understanding spatial entanglements is useful in exploring the interactions between the material and the human as well. The *boipara* can also be understood through Deleuze and Guattari's (1987) concept of assemblage, more specifically territorial assemblage. Territoriality in Deleuze and Guattari's sense points towards a relatedness between people, the spaces they inhabit and/or traverse, and the things that are also present in the space. Their account of the territorial assemblage provides ways of understanding how change occurs in assemblages not only through the intersections of bodies but in relations between people and their physical environments. I am looking to territoriality as a kind of door into questions of relationality between the various parts and participants of the *boipara* and spaces like it. However, notions of territory are also likely to prove useful in understanding belonging.

From such a position, I can never merely write what I have already experienced, but will always continue to think through and with those lived experiences, reshaping them, sending them towards new places. Regardless of what I may or may not have thought I was doing when I first chose to engage with narratives as a researcher and as a writer, I think I am using writing as a means to witness my own and others' movements in thought considerably more than I am using it as a productive alternative to 'tracings', lists or attempts to undertake ethnographic data collection. Narratives *must* make multiplicities because every narrative connects with those before it, around it and after it, and every narrative also relies on the many connections that writers and readers will make with a vast intertext and intertextual manoeuvres.

Materially speaking: Objects and things

In being a part of the *boipara* at different times, I have been able to produce a range of layered narratives through my interactions with the book buyers, sellers and other regulars. Previously, I have shown how I have navigated my own way, making my own sense of the *boipara* sometimes through the stories of my Baba and Dadu, sometimes through the directions pointed by Debashsish da the bookseller, and Poltu the tea stall owner, and through my wandering expeditions and personal visits to the space, both in the capacity of a researcher and an ex-regular. I have also become a part of the space through the many travels of my own second-hand books from and to the bookstalls over the years. Likewise, I also formed a long-standing relationship, modulated by different visits and familiarities, with the non-material components of the space.

As a former regular, I often realised during my fieldwork that College Street remains highly idiosyncratic in its functioning as an assemblage because there is a strong sense of simultaneity evident in what takes place there. There seem to be very few single or isolated events. Everything is, in one way or another, in relationship with everything else. In the information and stories I collected in my fieldwork, I began to notice that it was not simply the everyday intersections and interactions between regulars that made the *boipara* unique; the kinds of stories and memories that people carried with them informed their sense of the space and thus their interactions with others.

In my conversations, it became noticeable that we shared a collective pool of affective associations with the age of the books, the taste of the tea from the shanty tea stalls, the smoggy atmosphere inside the coffee shop. The everyday objects of the *boipara* participated in this way in meaning-making processes that were largely affective and sensorial in nature. Thus, what we shared was also *different*, because our meanings were formed from different experiences for different bodies at different times and could never be said to be exactly the same. More importantly, in my conversations with the regulars, those interactions between us presumably had different consequences, being joined to other stories and then others. In my case, the conversations became data, but that was only the first step.

As for the objects themselves:

> These days, you can read books on the pencil, the zipper, the toilet, the banana, the chair, the potato, the bowler hat. These days, history can unabashedly begin with things and with the senses by which we apprehend them; like a modernist poem, it begins in the street, with the smell 'of frying oil, shag tobacco and unwashed beer glasses'.
>
> (Brown, 2001, p. 2)

Indeed, within the *boipara*, the relationship I shared with the second-hand books that came to my house when I was a child, my becoming regular as a student, the benches I sat on when doing my fieldwork, the times I spent in the

Figure 5.1 The Indian coffee house, College Street

coffee house were more personal and affecting than many human interactions. I found something in the materiality of the forms and structures and objects that was strongly persuasive and insistent in relation to conjecturing the stories they would want to tell.

When objects are recognised as having their own stories, as carrying their own experiential narratives, they become *things*. Jonathan Lamb (2011, p. xi) points out that objects are important in social lives because of their capability to move. This mobility produces a 'commercial and symbolic value'; we study them as agents or actors (Latour, 2005) that produce and circulate social, economic, political and cultural knowledge. 'Things, on the other hand, are obstinately solitary, superficial, and self-evident, sometimes in flight, but not in our directions' (Lamb, 2011, p. xi). The latter is ironic in that Lamb admires the autonomy of things. There is a difference in how we form relationships with objects as opposed to things. Objects are primarily at the mercy of our subjective narratives. We use and interpret them insofar as they play significant roles in our cultural, social and political meaning-making processes. In this sense, objects are in equilibrium with us: their presence, although producing cultural or social knowledge, does not necessarily disturb the environment in which we find them. For example, let us take any book shop in Calcutta – say, the Oxford Bookstore in the posh, upscale Park Street,[1] in a recognisably upper-class area of the city. Park Street has a number of fine dining restaurants, jazz bars, exclusive tea and coffee houses like Flurrys and the Oxford Bookstore. The everyday functioning of this street is far removed from that of College Street. The Oxford Bookstore is emblematic of the city's elite. Characterised by interiors replete with glossy, shiny, new, *first-hand* books, placed in polished wooden racks and fitted with modern mood lights, the Oxford Bookstore provides its customers with a high-end, exclusive experience in buying books. The reading space inside the shop is the Cha Bar, a specialised tea-drinking, socialising and reading space. Here in the Oxford Bookstore, everything is organised, polished, new and custom-made. Browsing through its books, it is easy to form a clear idea of the kind of customers for whom the store is designed. The books tell us about the reading culture prevalent among the usual customers of this bookstore: the furniture, decor, the 'cha' bar, the mood lighting set on the interiors of the store is indicative of the class and everyday lifestyle practices not only of the people who visit this store but also of the Park Street area in general. The objects serve the purposes of marketing for high-end consumption: to be meaningful markers of a comfortable lifestyle so that they appeal to consumers who value such a lifestyle – whether they actually live it or only aspire to do so.

With things, it is slightly different. Things do not exist at the mercy of the subject's narrative and associations. The materiality of things assumes a distinctive autonomy that allows them to produce their own narratives. Drawing on *A Thousand Plateaus*, Bonta and Protevi (2004, p. 110) explain 'material' as

> 'molecularized' matter that displays traits linked to forces (capacities of self-ordering) (342–5). In other words, material is matter moved from equilibrium/steady state/stability to a far-from equilibrium 'intensive crisis' state and thereby able to manifest the effects of virtual singularities as it crosses thresholds (or, indeed, in certain cases to release a new set of singularities, new patterns and thresholds of behaviour).

The inclusive disjunctions involved here, which are also rhizomic relations, suggest that for Deleuze and Guattari (1987), as for Bill Brown (2001) and Jonathon Lamb (2011), the materiality of things matters because things can seem to have agency, or at least to act autonomously and carry strong potentials to disrupt the arrangements we try to make for them. I suggested above that we should resist giving the territorial assemblage itself actual *agency*, instead understanding it as having a degree of autonomy to the extent of it being the product of 'active, purposeful, effective arrangements'. I also observed that

> the assemblage in itself does not have agency, but it is the outcome of many choices and arrangements that have taken place and are taking place within it, and each of these has the potential to make other connections and linkages within and beyond the assemblage.

I want to suggest that, apart from human subjects exercising agency to produce the particular characteristics of a territorial assemblage, an important role in arrangements, movements, connections and linkages can also be attributed to *things*. The things that inhabit the territorial assemblage of the *boipara*, or travel to and from it, have been key to its emergence as an assemblage, the changes that have taken place and continue to take place within it and around it, and the permeability of its edges. Things *do* carry their own narratives and narrative potentials. For theorists such as Brown (2001) and Lamb (2011), this is what distinguishes them from mere objects that we can press into service. Things can be stubbornly resistant to being treated as objects and equally they can be particularly insistent on maintaining their status in relation to the more-than-human – for instance, in relation to the myths and mystique with which the *boipara* is invested by humans. To return for a moment to Massey's propositions about space and spatiality, it seems to me that things, and also objects-becoming-things, have a significant part to play in how the *boipara* unfolds, how it operates relationally, how it is always under construction and what I can glean from its 'stories so far' – a handful of which are 'mine', at least to the extent that any story can ever be 'mine' or 'yours' or 'theirs'. All my wanderings and wonderings, my physical and intellectual meandering through the *boipara* and through the ideas it raises for me, ask whether and how a space such as this offers fresh insights into spatiality and ways of working with it.

The physical area comprising the lanes, the by-lanes, the main road, the bookstalls and shops of the *boipara* – all of these have come to existence in an essentially uncontrolled, organic and sprawling fashion. In its everyday functioning, it is uncontrolled and can be understood as molecularised. There are no physical boundaries or barriers indicating where the *boipara* begins and ends. The space, in every direction, is becoming-*boipara* and thus also invites conceptualisation as becoming-spatiality. In this continuous movement, many of the objects in the space are carried towards *becoming-things*. This is enabled by the affective force of the language of the place, the innumerable conversations that occur there and the interactions that these accompany or describe or turn into stories. And while that movement is never still, never stabilised as a particular identity for the *boipara*, the relations between people, between people and space, and between people and things also express ongoing processes of various kinds of belonging. Sooner or later, for many regulars, those processes transform into *feeling at home*. In this, we can probably decipher a range of complex affective, material, social, individual-and-collective transformations that can also be understood as *becoming-insider*.

Our spatial memories are never limited only to human interactions. The composition of these scattered patches of memories that we remake through various processes of 'joining the dots' consists of much more. My experience of the interactions between histories, memories and the *boipara*, and similar experiences that others talked about with me, nearly always redirected individual memories towards collective memories, and collective memories towards a pool of associations. These associations are frequently with things, with objects-becoming-things and with the material world more generally – the built environment, sensory experiences of the precinct, comments about the 'atmosphere' of the *boipara* or the 'feeling' of it.

The *boipara*: Space and its relationship with language

The materiality of the College Street *boipara* is not limited to the physical objects found there. Every aspect of space has a material and a non-material component. Sometimes it is the words heard there; sometimes it is people's sense of the 'atmosphere of the place'. What stands out for me is that experiencing the *boipara* constitutes an intricate interweaving of the material and the immaterial. This is a complex intermingling of content and expression (Deleuze & Guattari, 1987, p. 88). It is perhaps impossible to describe or categorise everything contained in such a space. As I have stressed throughout, what exactly constitutes a spatial assemblage is not determinable, quantifiable or formulaic. It instead depends on subjective, affective experience. In attempting to be alert to the immaterial components and my affective responses to my encounters and observations in the *boipara*, what I began to notice was that – as with the objects – my reactions and feelings were shaped by the space itself. College Street (like all spaces, perhaps) is neither inert nor neutral in how someone like me engages with it. It allowed me and/or invited me to explore it in certain ways (even if those ways changed from day to day).

The *para*

One of the important things about the second-hand book market is its name. Although the *boipara* is an area mainly comprises book shops and bookstalls, publishing houses, local eateries and coffee houses, it was never just a market for us students, a bazaar or an area simply for transactions.[2] The area of the College Street second-hand book market was always called the *boipara*: it never acquired a name that in fact reflects *what goes on there* commercially or educationally. Further, for the regulars the word *boipara* has taken on a meaning associated much less with the commercial activities conducted there than with the idea of a space where people came to hang out and spend time while they browse and buy books. *Boipara* is a compound word consisting of two Bengali words joined together.[3] *Boi* translates as books and *para* means neighbourhood. *Para* as a part of Bengali language and as an idea of belonging plays a significant role in the everyday life of Calcutta. The *para* as the physical space of a neighbourhood and as somewhere a Bengali living in Calcutta wants to belong – neighbourhood as community – facilitates multifaceted interactions, material, affective, historical and sensorial.

The long stretch of main road that cuts through Mahatma Gandhi Road continues to be known as College Street until it assumes different names and points, both on the southern and northern ends.[4] Although often used to indicate the same place, College Street and the *boipara* are not quite the same thing conceptually or experientially. The usage of the word *para* can help us to understand the difference. Chaudhuri (2013, p. 118) writes that 'of all the institutions that distinguish the city of Calcutta, the most ubiquitous, yet in some respects most inevitable, is the *para*, the locality or neighbourhood'. The idea of the *para* is very closely associated with production of identities. It is never characterised by physical boundaries, but very distinctly associated with everyday cultural practices. The identities that a *para* generates are often complex because they are formed by numerous multi-layered associations. These associations are often grounded in everyday cultural narratives, practices and material associations.

The term *para* was never a part of the official administrative discourse of the city; however, it is the term with which people most identified. As Chaudhuri (2013, p. 119) further observes, in referring to areas of the city,

> [T]he colonial government preferred thanas (police stations) or the later system of wards, first instituted in 1852. Nevertheless, the paras remain an integral part of Bengali cultural memory, enshrined in language and indispensable to the everyday recognition of the city space.

It is important to note here how Chaudhuri recognises the enmeshing of memory and language in forming certain 'psycho-social' relationships with the city space. For her, Calcutta is a city that is not known for its *flaneurs* because the weather is not conducive to the act of walking and gentlemen did not choose to walk in the city.[5] This means the spatial associations that people made were often those 'psycho-social' attachments to which Chaudhuri refers, or what she also calls 'a domestication of social spheres' (Chaudhuri, 2013, p. 119). While I am in partial agreement with Chaudhuri's observation, I think our association with the *paras* can be explored further by going beyond her psycho-social reading.

Let me begin with my own personal understanding of the *para*. I have been familiar with the idea for as long as I can remember. As mentioned earlier, I grew up in Hatibagan, one of the oldest suburbs in North Calcutta. Despite growing up in the 1990s, around the time when rapid urban gentrification, 'the smart city' and mall culture began changing the cultural and geographical face of the city, even then North Calcutta remained quite protective about the *para* culture. In fact, the northern side of the city is the only part that continues to retain the word *para* to this day as an indispensable aspect of people's relationship with the everyday city and the language of its neighbourhood spaces. I think this retention is important in understanding the *para* in *boipara*. Every city space has its own language, whether as a way of talking or a set of words that is put to use in the everyday practices and lived experiences occurring there. As I have suggested earlier, this language is affective, sensorial and mobile (it has a history as well as a location) in addition to being literally a linguistic system. If I focus specifically on North Calcutta, where *paras* still exist and the second-hand book market continues as a vital factor in the local economy, the *idea* of the *para* remains an important part of the language that articulates the area.

It is from my experience of growing up at a time when the concept of the *para* was still being fiercely retained (while it was being rapidly lost in the rest of the city) that I realised how, as Chaudhuri (2013, p. 124) puts it, 'essentially, the para is a form of social practice, and it is a social practice born of a certain time. It cannot be willed into being, but it requires the exercise of collective will to survive.' But what constitutes this process of the 'collective will to survive'? As participants in the *para*, we always had certain things that we did there. The '*there*' of the para is its spatial composition. For me, the *para* included certain lanes and by-lanes that defined our neighbourhood. It always had a well-defined, material and affective meaning that came from a sense of belonging and placemaking in which I actively participated through everyday activities (as did any other person who has grown up in a *para*). Like my grandparents, parents, aunties and uncles, we had our own social circle in the *para*. These were *para 'r-bondhura* (friends of the neighbourhood). Spatially, it was – and still is – hard to pinpoint the boundaries of the *para*. However, affectively, the para always maintained its sense of boundaries:

> it recognised its own; as children, as visitors, we knew who was in and who was out, and so absolute was our sense of belonging that even today, meeting someone from that locality, we would describe ourselves as 'of the same para'
>
> (Chaudhuri, 2013, p. 121)

As kids and later young adults, we always had our own friends in the *para*, similar to the older members of my family. Put in these terms, it could be inferred that the *para* was very much a self-regulated and self-regulating space. However, what is interesting is that this sense of regulation was more often than not sensorial, affective and materially determined, in addition to (or indeed because of) its unique social processes and cultural activities. I will return to this point, as it will become crucial in my analysis of the spatiality of the second-hand book market as a *para*.

Being a part of a *para* also meant doing certain *things* that included, but was not restricted to, participating in certain activities. It is important here to stress the difference between *doing things* and participating in activities, particularly in relation to how people both used and inhabited the *para*. For example, most of the houses that existed along the lanes and by-lanes of my *para* had *ro'aks*. These cement extensions to the exteriors of the houses, around the entrances, were and remain typical of North Calcutta. Often, they look like cement blocks on which one could easily sit. This is where the *adda* of the *para* took place. It was a regular activity of my father and his *para* friends in their younger days, when they would to get together at the *ro'aks* every evening after they returned from college or university. On an everyday basis, the neighbourhood *adda* remain a very gendered practice in that they are male dominated and predominantly carried out at the *ro'aks*.

However, in my younger years, community events and activities often brought all members of the family – female and male – to become active participants in the *para*. This occurred, for example, during the time of Durga Pujo – the biggest socio-cultural and religious event every year in West Bengal, which lasted for five days. In the parts of Calcutta where *para* still exist, everyone who lives in the *para* takes responsibility for and participates in the celebrations.[6] Throughout the year, the *para* is a space for community events – whether sit-and-draw competitions or cricket or football tournaments. In these ways, the *para* remains the space where the neighbourhood materially, socially, culturally and affectively maintains and expresses its sense of community.

The workings of *para* are very similar in their material and affective interactions, irrespective of location.[7] While much has been written about the sociological and cultural practices of the *para* (see Chaudhuri, 2013; Chattopadhyay, 2012; Dasgupta, 1995), I am interested primarily in considering the material and affective associations we make with the *para*'s physical circumstances. This means returning to a focus on how the *boipara* is more than just a place where books are sold, publishing houses operate their businesses and people visit merely to hang out at the coffee house or study at the different education quarters in the precinct.

As a *para*, a neighbourhood, the *boipara* enables similar connections and responses, although the human participants have very different interests, and thus relationships. Materially, the College Street *boipara* has taken on a unique character that is far from anything that its colonial design and original purpose may have expressed. As the main road that cuts across the heart of central Calcutta, College Street is busy and congested. This is not unusual since Calcutta has never been known for smooth traffic flow and well-organised public spaces. The traffic on the main thoroughfare, Cornwallis Street, is perpetually chaotic, simply getting busier during peak hours. Similarly, the school, college and university students start to arrive in the College Street precinct between 9 and 10 am. Gradually, the bookstalls begin to organise themselves on the footpaths and begin their day. The book shops and publishing houses open a bit later and eventually the coffee house and other eateries also open for business.

Much like the functioning of my own *para*, there have always been *things to be done* in the *boipara*. The movements and rhythms of these activities draw both visitors and regulars into engagements that, on the one hand, may require the care involved in practice or, on the other, follow well-established habit. For many, though, the connection lies somewhere between, where practice is becoming automatic and habit has not quite forgotten intention and awareness. The threshold of becoming-a-regular, this also identifies the underlying logic of production of the spatial assemblage since these activities and interactions are intrinsically material. As a physical space, the *boipara* is affectively perceived through connections with, and connections within, its materiality and its material objects. I have noted already that every *para* is characterised by the idea of *things that one does there*. My own early connections to College Street – even though I had not yet been there – were shaped by a belief that everyone who went there did so for a purpose, to do something specific. As also described above, when I began visiting the *boipara*, I went through a phase of self-awareness and reflexivity about my presence there. My recognitions were about both human and material interactions, and the affective connections I made with the physical elements of the space were as important as those I made with the people. Sometimes these connections were tied to others, but mostly the relationships were autonomous.

Thus, I treasured my visits to the coffee house for the regular meetings and the coffees and conversations I shared with my friends. But on many occasions I went to the coffee house just for the simple pleasure of being there, sitting among the wooden tables and chairs, beneath the skylight windows that illuminated the defining material features of the space. What stood out and later became the central aspect of my experiential knowledge about the *boipara* as well as the coffee house was that sense of 'material'. The following excerpt from a fieldwork conversation is a good place to start exploring this sense.

Interview and conversation with Mainak, September 2016

Me: How long have you been taking care of this bookstall?
Bookstall owner, Mainak: Since the 1960s.
Me: You must have seen the space change drastically since then?
Mainak: College Street never really changes [smiles, conveying a sense of pride]. Yes, the old set of second-hand books are replaced by a new set of second-hand books. Over the years, the old regulars remain and thrive while new regulars come to become a part of the space. However, to answer your question, I often think of the changing streetlights over the years to describe the passing of time in the boipara.

In the 1960s when we first set up the stall, we had candles as the main source of light. It was all good during daytime because of the sunlight, but as afternoon approached, we used wax candles. It was fairly dangerous, you know, old books and wax candles are perhaps best when not kept in close proximity. Then the gaslights came and that made our lives much easier: the regulars of my shop, who often sat inside on this wooden bench looking for books or even reading them in the stall, found it much easier. In the later years – the 1970s and 1980s – the collective force of the gas lanterns and the streetlights from the lampposts on the footpath made reading in the evenings better than ever. Mind you, until about the later 1980s we still had a lamplighter who had to come to the streets regularly to turn on the streetlights on the footpath. On the days when these lamplighters decided to give our lane a miss or simply forgot (although this was a scarce case, as the boipara *is around one of the primary city junctions), we often got back to our old days of depending solely on candlelight.*

Over the 1990s the streetlights changed for the better and so has the lighting system inside our stalls. You will notice that all the stalls have light bulbs or tube lights now. However, I still have regulars from the 60s casually mention how much fun it was when we just had candlelight and gaslight reading, reminiscing [about] their past experiences under different and far-removed lighting conditions. This makes me wonder, how has this place changed? How do we perceive change? What I find more interesting is how the young regulars of these days often come and say how they would have preferred the candlelight and the gaslights over what they perceive to be the 'over-lit' street these days. I often ask my regulars why they don't spend as much time as before, to which they reply how the present atmosphere tempts them less. I have often tried to reason with them: the lighting is better now and that it should make their reading experience much smoother. On the other hand, I have these youngsters, the college-going regulars of today, come and ask me why do I not have the gas lanterns and the candles inside the stalls anymore. They want to be able to experience the atmospherics of the boipara *of the 1970s. The still existing* ro'ak *inside the by-lanes, the coffee house, Putiram and a few other old joints are their refuge. Some of the bigger book shops still have the old three-legged uncomfortable wooden stools from the British era; the coffee house still has the squared wooden tables and chairs; the skylight still acts as the primary source of light in the daytime.*

There are a few moments in this excerpt that are worth pondering. My question to the bookstall owner was primarily intended to invite him to tell me about his experience of changes in the *boipara* over time. I wondered why he chose to trace these changes through the history of streetlights on the footpath and lighting in his stall. Given that he observes that 'College Street never really changes', the most significant or obvious changes for him and not in the structure or ways of using the space *per se* but in its atmosphere as a result of the changing sources of light available in the city. That is, the effect of the material on everyday experiences, and affective responses to that, stood out for him more than any shifts in human interactions that might have taken place, such as customer behaviours and expectations of service or stock, differences in reading habits or more general differences in how people used the space. It is evident through his thoughts and expression that the change in the atmospherics of the space because of changing sources of city lighting is what was most prominent in his assessment of transformations in the *boipara*. His observations point towards a sense of collective affect through the part played by material things in the production of everyday spatial experiences. The materiality of the *boipara* is, for him, the source of his primary sense of its spatiality. In his business, light makes all other experiences possible, including the owner's interactions with customers who would come to buy, sell or browse through books, and their interactions with each other and with the space itself, the cramped stall and the single wooden bench. In turn, light facilitates the affective resonances borne by all those contacts. Most of these we can only guess at, but light has its own affects, as the nostalgia – both real and received – for the candles and gaslights on the part of many older and current customers indicates.

But if the material aspects of changes in the *boipara* were the most obvious to Mainak, he was also aware of the human side of things. Later in the conversation, he mentioned how the old habit of the regulars of his stall coming and reading inside the shop had changed: they spend less time there now. I wondered whether the change was because people had just become busy or lost touch with the past affective registers of the space. He was quick to point out that while people's investments in certain spaces might change, this was unlikely in the case of the College Street *boipara*. He pointed out that his regulars often mentioned how they missed the atmosphere of the stalls in the old days. Thus – as indicated by his recognition of the importance of the regulars' expressions of and understandings about connections between nostalgia, reminiscence, and 'atmosphere' – even when material changes mean that things must be done differently and are able to be managed more conveniently, people can and do choose to retain their affective connections with a space. Although Mainak did not itemise particular aspects of other material changes to his stall and the *boipara* that he thought were important to his long-term customers, in addition to specific affective relationships with the object world, their nostalgia reveals what I discuss above as a feeling of being at home [*chez soi*] (Bonta & Protevi, 2004, p. 158). This is surely an indication that affective responses to materiality, to things and to objects-becoming-things is key to the creation and perpetuation of a sense of belonging.

My question here directly centred the bookseller as subject since I asked him how *he* had witnessed a change. To my surprise, he stripped off the privilege of centring his own subjectivity, and instead focused on the changing materiality involved in the lights. He also describes an experience that is about changing *materialness* through lights, and the affective influences of this on the others using the space. His response derailed any expectations that the conversation would proceed in a straightforward direction. Despite my asking a question that presumed a subjective reply, Mainak responded to what Thrift (2007, p. 8) calls a *material schematism*, with which he was familiar due to his many years working in the *boipara*. In discussing the second tenet of non-representational theory, Thrift (2007, p. 7) emphasises the importance of being 'anti-biographical' and 'pre-individual'. Instead, he conceives of a 'material schematism in which the world is made up of all kinds of things brought into relation with one another by many and various spaces through a continuous and largely involuntary process of encounter, and the violent training that such encounter forces' (Thrift, 2007, p. 8). The *boipara* is to a very great extent such a space where things and subjects come together through an 'involuntary process of encounter'. In doing so, each of these components creates its own narratives within the wider environment of the *boipara*. This is a process in which the material does not remain in the background and thus foreground the subject. Instead, it is the material that drives the spatio-temporal narrative. Thus, to the bookseller, change is understood as having taken place through material changes in street lighting and thus stall lighting, and the sensorial registers of the presence and absence of lighting in the street.

Discussions in cultural geography have frequently raised questions around the issues of home, belonging and materialising geography. These questions converge significantly on how we think of material cultural and spatial practices. The focus towards which the bookstall owner points us suggests that we need to attend to Jane Bennett's (2010, p. viii) guiding question at the beginning of *Vibrant Matter*. She asks: 'How would political responses to public problems change were we to take seriously the vitality of (nonhuman) bodies?' Although I am not addressing 'public problems' explicitly, that sense of the vitality of the non-human is a crucial recognition in relation to my primary motivation to explore the ways in which we can think and write about city-spaces and public spaces such as the *boipara* in more inclusive and subtle terms. From this, we can glean some important understandings for reading the *boipara* as assemblage. Since the material elements of that space circulate in affective relationships with the human participants – regular and occasional, long term and recent – and because these relationships are developed through the gradual opening of practice onto belonging (feeling at home) and then habit, we are drawn into complex arrangements that cannot be exhausted simply by seeking to decipher, enumerate and describe such a multitude of elements. Once again, this indicates that an assemblage is a point of departure.

From the footpaths

The footpaths play a crucial role in the unfolding of the *boipara* as a *para*. They also perform as spatial intensities – zones of close interaction between the regular subjects and objects of the space. In this sense, they act as catalysts for the circulation of a collection of affects. There is a unique arrangement to the footpaths of College Street. Along the inner side of the five-foot-wide strip are the university colleges and buildings, the coffee house, the book shops, the publishing houses, the buildings and printing and publishing businesses. Facing these buildings on the outer edges of the footpath is the almost continuous chain of makeshift bookstalls of changing shapes, sizes and colours.

The bookstalls back onto the traffic and face towards the buildings housing the established bookstores and publishers, turning the footpath into a kind of arcade, a passageway that, with all its diversions between the multiple rows of second-hand books, becomes a kind of maze following patterns that have been there for generations. Walking there involves continually navigating sensorial, affective information, repeating the rhythms and resonances that characterise the space while responding with different emotions to one or other or many of the intersections that occur. As a regular user, Paromita describes her affinity with the *boipara*.

Figure 5.2 Booksellers who often perch in between the books within makeshift bookstalls

Paromita's remarks in interview/conversation, August 2016

I am an avid reader and collector of second-hand books. I finished all of my school and university education depending on second-hand books. It was a matter of affordability then. It has become an addiction. Our financial conditions then did not support me to buy new books to pursue my education. Thus, I became familiar with the boipara *from a very young age. Look, I know every nook and corner of this place. This feels like my second home. The yellow musty pages of these old books have been with me all my life. I went to a school where not many kids used second-hand books. Most of the kids could afford new books. I like sitting on this bench on the footpath and reading while browsing through what I want.*

Like the bookseller Mainak's comments on his customers' responses to lighting, this interviewee's comments suggest that the people who are part of this space do not merely experience affective responses moment by moment, but think through them – especially with changes that take place over time.

Brian Massumi (2015, p. vii) observes that 'thinking through affect, is not just reflecting on it. It is thought taking the plunge, consenting to ride the waves of affect, on a crest of words, drenched to the conceptual bone in the fineness of its spray.' Both Mainak and Paromita reflect on the *boipara* in ways that are not confined to reporting on episodes or describing situations but extend into personal interactions and connections – whether their own or those of others – and the reactions and feelings that accompany these. I have noticed much the same thing with my own recollections when I have tried to record them for this thesis. Our narratives are not linear, and they spread into different dimensions in responding to the complexities of the space and so begin to resonate with the kind of movement in affect to which Massumi points. To reduce an account of a space such as the College Street *boipara* to mere history – to stories about interesting characters or significant events – is insufficient.

Thus the bookseller's story about the changing technologies of lighting in College Street emerges through details about his own stall and the people who have come there over the years, and the past that they construct through their responses to and feelings about lighting; in turn, the relationships with lighting produce relations with the second-hand books (which, if I were to extend this assemblage, would then be recognised for the exchanges and circulations to which they are connected). There are two important conclusions to be drawn from this. First, in charting the spatial assemblage of the *boipara*, it is not so much the subject (the human) or the object (the material, non-human) that matters, but the affective resonances produced between them. Second, it is also evident that 'affect is only understood as enacted' (Massumi, 2015, p. vii).

Therefore, affect becomes personal-and-political, and plays a crucial role in understanding the space as a concept. This political dimension to affect is also useful to my developing ideas about assemblage as a conceptual tool. I have established that our experiences of being (ontology) in the *boipara*, understanding (epistemology) and doing (activating) the space are characterised by mobility, relationality, connections and change. In realising that we have affective connections with the space of the *boipara*, we come to *think through* them, and we act in them in response to the richness of those thought processes, thus continually negotiating and creating a space full of 'potentiality or possibility' (Anderson, 2006, p. 733). Similarly, as a conceptual tool, assemblage facilitates the moment of thought every now and then, which then propels a movement towards that which is unknown but possible.

Notes

1 The store has a website that provides a virtual tour of all its shiny attractions. The Oxford Bookstore is testament to the fact that objects are fashioned in relation to us and our expectations and identifications.

2 Although Turner (1982) refers to a bazaar as a liminoid zone, it is nevertheless characterised and controlled by heavily capitalistic ideas of transaction and commodification.

3 Both separately and together, they make sense. However, in the context of my project, the word *boipara* encompasses several ideas, affectively, materially and historically that capacitate and activate the space of the *boipara* to function as an assemblage.
4 At the northern end, College Street becomes Bidhan Sarani and at the southern end it becomes Nirmal Chandra Street, which then becomes a part of a new neighbourhood, Boubazar.
5 The act of walking was heavily embedded in ideas of class in colonial as well as post-colonial Calcutta. The *bhadralok*, *babus* or the gentlemen did not choose to walk. It is hard to pinpoint these groups of people within the traditional understanding of class.
6 During these days, the city celebrates the homecoming of Goddess Durga after she destroys everything evil in the world. Every *para* in the city erects its own *pandals* (which are now massive installations), where everyone in the specific *para* is actively involved in the organising committee. The sociological unfolding of this event is emblematic of the coming together of everyone in a neighbourhood, or a *para*, as a community.
7 We also cannot ignore the fact that the *para* culture is rapidly moving towards being extinct as a spatial concept, even in the northern parts of the city. However, the College Street *boipara* remains and continues to thrive in its everyday practices.

References

Anderson, B. (2006). Becoming and being hopeful: Towards a theory of affect. *Environment and Planning D: Space and Society*, *24*(5), 732–52. https://doi.org/10.1068/d393t

Bennett, J. (2010). *Vibrant matter: A political ecology of things*. Duke University Press.

Bonta, M., & Protevi, J. (2004). *Deleuze and geophilosophy: A guide and glossary*. Edinburgh University Press.

Brown, B. (2001). Thing theory. *Critical Inquiry*, *28*(1), 1–24. https://doi.org/10.1086/449030

Chattopadhyay, S. (2012). *Unlearning the city: Infrastructure in a new optical field*. University of Minnesota Press.

Chaudhuri, S. (Ed.) (2013). *Calcutta: The living city vol. II: The present and the future*. Oxford University Press.

Dasgupta, K. (1995). *A city away from home: The mapping of Calcutta*. University of Minnesota Press.

Deleuze, G., & Guattari, F. (1987). *A thousand plateaus: Capitalism and schizophrenia*. University of Minnesota Press.

Lamb, J. (2011). *The things things say*. Princeton University Press.

Latour, B. (2005). *Re-assembling the social: An introduction to actor network theory*. Oxford University Press.

Lorimer, H. (2005). The busyness of being 'more-than-representational'. *Progress in Human Geography*, *29*(1), 83–94. https://doi.org/10.1191/0309132505ph531pr

Massumi, B. (2015). *Politics of affect*. Polity Press.

Thrift, N. (2007). *Non-representational theory: Space, politics, affect*. Routledge.

Turner, V. (1982). *From ritual to theatre*. Performing Arts Publications.

6 The *boipara* as an event

The *boipara* as an eventful heritage space

The *boipara* is never static, particularly because of the movements of people and books, and the ways in which the books are arrayed and the people interact with each other. My writing needed to be adequate to all that movement simply in order to 'keep up with it', to participate in the ebb and flow, to understand it through my own interactions and my own recognitions. I have not attempted to write *only* in and about the *boipara*, confining myself to events and conversations that occurred there. Like the books themselves, we all have lives outside the *boipara*; we all circulate through events and interactions in many other places and we carry knowledge and experience, as well as feelings of many sorts, with us wherever we go. So, in writing this book, I have tried to produce a kind of fabric from threads of personal stories and experiences, the stories of others and the experiences they have passed on to me, the memories and affections that have told me something about how family, friends, fieldwork participants, people I have met elsewhere or whose words I have read, understand and respond to College Street. That is, I have woven an existing pile of personal narratives that were a part of my own experiences together with the stories, experiences, narratives and analyses I have encountered. When stories and impressions and conversations and movements and experiences and interactions and sensations and objects and the feelings they provoke, and anything else that might be discovered about a place, are brought together in a 'common ground' of thought (no matter where they came from and how they were collected) there are different ways of making sense of their coexistence.

To invoke Deleuze and Guattari (1987), there are different ways of encountering the assemblage from which they emerge, in which they participate in and from which they depart. One way is to find whatever it is they share – to stabilise any problems, complications and uncertainties through similarity. Another, more difficult but much more productive approach is to look for how their differences communicate with each other, rub up against each other, resonate or fall into dissonances that still hold them together. Difference always remains uncertain, unstable and, as Clare Colebrook (2002, p. xxv) points out, 'any connection also enables a line of flight'.

DOI: 10.4324/9781003293026-6

Lines of flight can be of two kinds: relative lines of flight and absolute lines of flight. 'A relative line of flight is a vector of escape, a move between milieus' (Bonta & Protevi, 2004, p. 106). An absolute line of flight, on the other hand, is 'an absolute deterritorialization to the plane of consistency, the creation of new attractors and bifurcators, new patterns and thresholds' (Bonta & Protevi, 2004, p. 106). Within the spatiality of the *boipara*, we can discern both relative and absolute lines of flight. To do so might require a brief detour to revisit earlier subjects. In Chapter 5, attending to the potential affective relations between people and the material circumstances of the *boipara*, I provided an excerpt from my interview with the bookseller, Mainak, who responded to my question about changes in the quarter with a history of changes to street lighting in the *boipara* and the effect on the illumination of the stalls, which disturbed many customers. Thinking about the interior of the stalls opens up another recognition about how the *boipara* functions.

Mainak's own stall looks like an enclosed box made of tin, with an asbestos shade. It is difficult to estimate the size of the stall, given the piles of books everywhere. As I have described throughout, that is the same for all the stalls, which vary in size but rarely in their degree of clutter. Whether from outside or inside, the stalls look and feel claustrophobic. They are enclosed on each side by makeshift tin walls, but piles of books at the front also function as a kind of boundary, which also joins what is almost a continuum of such frontages with the hundreds of neighbouring stalls. The entry to a stall is usually along one of the walls. As already described, inside there is always a dilapidated wooden bench and sometimes a stool. The stool is often occupied by the bookstall owner, whereas the wooden bench is for the regular customers to sit on and read.

During my fieldwork interview and observation process, I sat on one of these stools at Rakesh da's shop. A number of times in our conversation Rakesh da mentioned that he wished he had a slightly bigger stall, so more people could come in. A bigger stall would have meant more space for shelves along the walls, and thus more space for books. Interestingly, though, Rakesh da never expressed any wish to own one of the book shops in the old colonial buildings. It was as if he was imaginatively and bodily inseparable from the idea of the bookstall: all he needed was more room for more books. Perhaps affirming his own attachment to his stall was the attitude of his customers. He returned again and again to stories of his customers' enthusiasm for coming into the stall, browsing for books and sitting for a long time on the bench, sometimes to decide on which books to buy, sometimes just to read for its own sake.

The uncontrolled, overlapping series of bookstalls spreading onto the footpaths of the *boipara* imparts an idiosyncratic mixture of connection and openness to the space. However, the inside of these stalls is quite the opposite: they usually feel enclosed, hemmed in and congested. Only one or two customers can fit inside each stall at a time, which means – as is often the case – that several other people might be standing outside waiting to get in when things

Figure 6.1 Books organised on footpaths outside the bookshops

get busy. They find a spot wherever they can on the footpath, sometimes leaning on a street lamp or the outer railings of the footpath, reading whatever books they can get their hands on and hoping for the crowd to thin, or pressing the stall owner to find this or that book or other materials, thus adding to a curious impression of systematic mayhem. But the regular customers never seem to complain about the lack of space.

There is much that is relatively static on a daily basis in the *boipara*: the attitudes and behaviours of the stall owners, which follow regular patterns and expectations, and the material composition of the bookstalls, the walls, the shelving and the books themselves (though many of these disappear and reappear every day). But there is also constant movement: the comings and goings of people in the streets, the various commotions that I have just mentioned accumulating outside one or another bookstall, people exchanging money and books and other materials, or exchanging ideas in the coffee house. The bookstalls spread and sprawl across and along footpaths whenever any additional space becomes available. The cars, bikes and trams rattle past on the road alongside. Everything is in a kind of flux, which runs through, not outside and separate to, all those other things that are largely static and unchanging.

In all this, there is one thing that is clearly both stable and in motion: the book. Of course, there may be more than one object or characteristic that is stable and in motion at the same time, but the object that matters to this book is the book itself. The book in the *boipara* (in a way that is not the case for a book in a book shop or on someone's shelf at home) is a point of contact between movement and fixity, a point where these recognise each other, opening up the possibility of a 'vector of escape' (Bonta & Protevi, 2004, p. 106) or a line of flight. But before exploring this more deeply, it is useful to notice some other aspects of the connections and movements of these second-hand books.

Books constantly come and go to and from the *boipara*. I have referred to this as circulation – and books certainly return and depart often enough to think of their movements as 'circular'. They can, however, also be thought about as rhizomic. It is important to recall that 'any point of a rhizome can be connected to anything other and must be' (Deleuze & Guattari, 1987, p. 7) – that is, rhizomes always return to themselves as well as making connections beyond themselves that may prompt new rhizomes. Books purchased in the *boipara* generate rhizomic connections between the *boipara* and the world outside the *boipara* but, given that books can return to any stall, their comings and goings generate rhizomic connections between different parts of the *boipara*. It may also be that they will become part of rhizomic connections between different individuals and households outside the *boipara* before they return for resale, perhaps by someone other than the person who initially took them from the *boipara*.

Books may never return, so creating a rupture in the tangle of interconnectedness that I have just described, but this does not mean the book in question ceases to have the capacity to carry the rhizomics of *boipara* further into the world and across time. Any rupture is an opportunity for the rhizome to 'start up again on one of its old lines, or on new lines' (Deleuze & Guattari, 1987, p. 9). To reflect on the many connections, disconnections, reconnections, multiplicities and trajectories involved in the rhizomic movements generated from the *milieu* (as middle, as in-between) that is the *boipara* is to understand how assemblages form and are never closed, always involved in movements beyond themselves.

When books return to the *boipara*, as I discussed in detail earlier, they return *changed* – sometimes materially with pages torn or dog-eared, sometimes carrying the traces of other readers' thoughts and imaginings in the form of marginal comments, clever observations or puzzled questions, sheets of notepaper forgotten in between the pages. The book's virtual movements are even more rhizomic that its actual movements: the ideas it contains, its illustrations, the stories it tells and the additions that accidentally or deliberately travel with it enter into new rhizomic connections and other possible arrangements – other assemblages, even creating new assemblages, which collect together other ideas, other people, other experiences in seminar rooms or coffee houses, changing

Figure 6.2 Booksellers organising the bookstalls

what people think, how they feel, how they might engage with someone else altogether. These connections also weigh on the book in the other direction, carrying all the potential of movement back to a stall in the *boipara* and to the pile of other books, with their own recollections of movements, waiting to be purchased again.

Books that find their way to the *boipara*, and spend years coming and going, to and from it, become, in a sense, hyper-capable of fulfilling these functions of '*the* book'. The capacity of books to form rhizomes 'with the world' is intensified and re-intensified again and again by the *boipara* itself, the stall owners, their customers, the *adda*, the additions to the books that accumulate over time as a result of the books' movements through many spaces and many hands, and so on. The sense of intensity is surely part of the appeal of the *boipara*, its 'special' atmosphere, its role in cultural and political transformations.

The *boipara* as I have explored it throughout this book consists of multiple movements and connections and opportunities for intensity or 'speed' to build in any part of it at any time. It actually *relies* upon perpetual movements of departure in the form of books and the people who read them. As a productive assemblage, however, we have seen that the *boipara* thrives as a space with particular cultural significance, not only because of these multiple but obvious movements inherent to the exchange of second-hand books but also because of all kinds of intellectual, cultural and political exchanges – exchanges between this part of the city and other parts of the city, Calcutta and the rest of the West Bengal, Bengali culture and the nation, and the constant exchanges between past, present and future. Above all, the *boipara* is involved in the rhizomics of affect and imagination – which, as discussed above, are a matter of intensities. For Deleuze and Guattari (1987),

> [T]he degree of intensity that builds in any assemblage is what is most likely to contribute to the gradual or sudden emergence of fissures, ruptures, escapes, departures – that is, lines of flight – from that assemblage along new trajectories that may produce new rhizomes, new assemblages, new intensities, new ruptures, new lines of flight.

A line of flight of the sort most often prompted by the *boipara* creates an oscillation between questions around 'what used to happen' and 'what can happen' enabling a narrative of imagination and anticipation, a 'vector of freedom, or at least freedom-from' (Bonta & Protevi, 2004, p. 107). It is, in Deleuze and Guattari's (1987) terms, a relative line of flight, not an absolute one. Since we are not able to trace its movement, we might even have to call it a provisional line of flight, but it is a circumstance in which fixity and movement connect with each other, exchange values and open up new kinds of stories about the *boipara*. In the following paragraphs, I explore how these departures create eventful moments or intensities within the *boipara*.

Towards the event: Intensities

In Chapter 5, I discussed the different parts and participants that play vital and active roles in relation to the overall sensorial and affective experience of the *boipara*. Any time spent there was to experience a soundscape made up of trams moving through the cobbled streets, the voices of the booksellers incessantly calling '*didi, didi*',[1] the noises of the printing machines in the publishing houses, the countless conversations in the street and in the bookstalls and the coffee house. It was also a visual spectacle of people and things moving, the colours of clothing, the colours of the walls, the colours of the books. Then there were the smells, of people and cars and cigarettes, of coffee and the smell of food drifting from the coffee house too, or drifting in from elsewhere. Everything happens simultaneously, like an orchestra, but not in any way harmoniously – it is less 'in harmony' than like an orchestra preparing to play, tuning up, with everyone concentrating on what they are doing individually or in their particular small groups.

In many instances throughout this book, it has become apparent how in the *boipara* the second-hand books never remain *just* second-hand books; the coffee house is never *just* the coffee house; the interactions, conversations, walking and sitting inside the shanty bookstalls never remain *just* that. They may have ordinary names, nouns that 'contain' an aggregate of references and qualities and so on, but that is not all they are. The have become loaded with all sorts of other associations and values. Yet they cannot be pinned down as unique, allowing a description to exhaust the object or activity. Many of the objects and interactions to which I have referred to are common, everyday – regular but also irregular – spatial practices that take place in all sorts of other situations. People do not exclusively read second-hand books in the *boipara*; they do not smoke cigarettes just in the coffee house, or even just in coffee houses; the *boipara* is not the only place where people strike up casual public conversations, go strolling down memory lane or solve the problems of the world through their political discussions and debates. These are routine activities and everyday spatial practices that unfold in many urban settings. So why is it so difficult to think of and write about the *boipara* as 'just like' anywhere else? It is because that is not really the question. It is not about repetition and the order of the same.

My work is trying to tease out the hidden intensities and 'far-from equilibrium' processes that lead to the *boipara* being in many ways a space that *never* settles, that resists stability, and is thus particularly productive of rhizomes, multiplicities, assemblages and lines of flight that carry it elsewhere. For this part of my discussion, what matters is the recognition that representation pursues the stable, extensive (material) characteristics of objects and the stable, stratified (institutionalised) arrangements between people rather than seeking the forces (thermodynamic or electromagnetic, for example) and power relations that produce objects and structures and regulations. How can we represent a force or the workings of power, after all, except through their effects on things, or except through how they act on bodies, its affects?

This is why I have questioned the inclination towards representationalism that comes with writing. The constant movement of the *boipara* and the sense in which what seems stable and fixed is also always transforming, suggest that there is more to recording encounters there than can be achieved in historical, cultural or political representations of the city space, and that it is important to find ways to point towards the intensities and forces at work there, which can be recognised in the affective responses of the booksellers, customers and, in my case, researchers who spend time there. There is, inescapably, a kind of representationalism involved in it, so I have therefore tried to account for my own interactions with the *boipara*, following the trajectories of my own lived experiences of the place and the connections (as well as affects) these can disclose, in whatever ways seem likely to be most revealing, depending on what I am focusing on. This has been an attempt to make connections, to build an assemblage that might in turn discover/uncover flows, intensities, forms of escape, lines of flight towards other understandings as well as other assemblages.

There is no denying that my thinking about what I have encountered, listened to and recorded has relied on historical, political, social and cultural writings and that, in a sense, I have thus produced my own representations of these – just as I have produced my own representations of the various accounts, stories, memories, interactions and descriptions given to me by others from their own perspectives in relation to the historical, political, social and cultural backgrounds against which they live their lives, go about their business, develop their own knowledge. But again, how I have reflected and reflected on all this is not an ethnographic effort to represent the stories and memories and ideas of others. I have instead noticed the rhizomes that did (or did not) form in and from the connections I made with those people and ideas, and joined their stories and observations to the growing assemblage of my writing. I hope that, as a form of expression of my experience, insofar as it still relies on representation, this book has become a collection of, to borrow Anderson's (2018, p. 3) term, 'representations-in-relations'.

In avoiding a systematic or cohesive account of my own experiences of the *boipara*, I have focused on three periods: discovering the idea of the *boipara* (the place that existed for me in the stories and books belonging to my family); my time as a student; and my research – the field trips, observations and interviews that produced a version of the space that circulated between my own academic observations and the stories told to me by others. Remembering and recording all this, even if it is really no more than a series of fragments, has often produced affective responses that, in turn, have become connected to and extended the assemblage of my interactions with the *boipara*, just as whatever I have learnt (reflexively and from others) has extended the terms of that other, connected spatial assemblage: the College Street *boipara* itself. Of course, that assemblage (as is the case with any assemblage) opens onto other assemblages, in relations with the comings and goings of the booksellers, all those who visit here regularly or occasionally, the conversations in the coffee house and wherever they carry to and whoever they influence. I have only been able to glimpse these in the memories and the observations and the stories told me by others, and I have only been able to make some provisional suggestions about the assemblages still continuing to make their connections somewhere else.

The event and everyday spatial practices

In the tenth series of *The Logic of Sense*, Deleuze (1990, p. 63) writes:

> Just as the present measures the temporal realisation of the event – that is, its incarnation in the depth of acting bodies and its incorporation in a state of affairs – the event in turn, in its impassibility, and impenetrability, has no present. It rather retreats and advances in two directions at once, being the perpetual object of a double question: What is going to happen? What has just happened? The agonizing aspect of the pure event is that it is always and at the same time something which has just happened and something about to happen; never something which is happening.

My interest is in looking at how the idea might allow us to think about and understand the *boipara* not as a 'site', but as a place/space of constant movement, interaction and change. With the idea of the event, Deleuze is effectively pointing towards understanding how something happens. Given the multiplicity of incidents and interactions and situations that it is possible to witness in College Street, and given how quickly one incident becomes or is replaced by another and another, the event as a process and concept helps me to recognise something very significant about how life in the *boipara* takes place. Events are not simply *there*, as if they might be captured by a camera – that is, as if they are accessible to the processes of representation (as bodies, objects and so on are taken to be). The event is only recognisable in its effects and affects, in the traces of its coming to be and its disappearing. That is why, in talking about the *boipara*, I have chosen to emphasise words such as folding and unfolding.[2] And so the multiple ways in which we can experience the everyday unfoldings of the College Street space are always characterised by a sense of oscillation, of mobility or flux. The present as a static, representable moment in time is never really available: it is about to happen/it has just happened. When we try to register this sort of rhythm in our sense of experiencing the *boipara*, experiencing any space, our present is continually caught between the past (memory, whether immediate, recent or distant) and the future (anticipation, imagination). That betweenness opens the realm of affect in a movement that we can really only recognise in another oscillation: between Deleuze's questions 'What has just happened?' and 'What is going to happen?'[3] For Bonta and Protevi (2004, p. 101), it is the *dissipated energy* that is produced because of this movement or vibration that characterises the congealment or the 'intensive' or 'far-from-equilibrium' state. Without equilibrium, this 'state' is neither static nor settled: events produce other events, propelling changes in the assemblage, opening towards other events.

As Steven Shaviro (2007, p. 1) observes, the world is 'made of events, and nothing but events: happenings rather than things, verbs rather than nouns, processes rather than substances'. The point I need to make about this is that events are not just things that happen: they happen *to* us (or to others) and we can also impel and participate in them. Our everyday spatial experiences, including moving, walking, feeling, interacting and even writing, are all part of 'rhizomatic multiplicities' (Bonta & Protevi, 2004, p. 101) that effectuate the event within the space and beyond. This reinforces what I have suggested already: that writing does not represent the event (which is impossible anyway), but participates in it. Writing is best thought of as an event itself, joined to other events rhizomically, opening towards other assemblages. In the context of the *boipara*, it is writing – or, more broadly, communication: conversations, stories, debates – that connects with more or less everything that happens in the *boipara*. Writing thus becomes the possibility along which or from which so many lines of flight might begin to run – relative lines of flight, mainly, and some with a trajectory that passes only from one part of the space to another while others, as I have mentioned with the movements of the second-hand books, may pass towards an outside, making new connections and forming new assemblages as they go.

The bookstalls and the books shops have varying times of opening their shops for business, usually between 10 and 11 am. However, each bookstall and book shop and publishing house has its unique routine, which is followed every day. The working hours of the *boipara* perfectly coincide with the working hours of the surrounding colleges and universities, which makes perfect sense. Once open and ready for business, the bookstall owners often act as 'curators' of the space, helping new students to find their way around. They will also help a student find the books required for a university syllabus or engage with other customers in conversations about the day's news or politics or union activities – conversations that are often left unfinished and sometimes picked up again days later if the customer happens to pass by again.

Interview comments from Bookseller Quasim

Initially I had a second-hand bookstall only; however, over time I focused on keeping copies of syllabus and question banks,[4] *government and civil service exam forms along with my second-hand books. This also seems like a useful thing to do to bring new students to the bookstalls. Even coming into one [is a start].*

I asked him how this happened. He explained:

> *Every student who has enrolled to the universities has to sit for exams. They will inevitably need question banks, syllabus books, last five years suggestion papers etc. We collect and compile them regularly and make them accessible to the students. Over time (sometimes weeks, sometimes in the first few months of becoming a regular of the space), they start to come in by themselves. Gradually their interests shift from just getting what they need to browsing through books, etc. Not everyone becomes interested in books, notes, etc. but some of them certainly do. Just the other day we had someone who, for a long time, only came to our stall to get her curriculum books. That is, during the time when her semester was running she was only buying books, or loaning them from us, for the purposes of her exams. Now that the exam season is over, and the universities are closed, I was surprised to see her. She walked into the stall quite awkwardly, not knowing what she wanted. I asked what she was looking for. To which she replied that she had always wanted to come to the* boipara, *with some time on her hands, where she would be able to browse through the books, sit and spend time in the area and buy books that she enjoyed reading. She confessed that she is not an avid reader and is not into buying books and reading for pleasure regularly. However, after three years of undergraduate university life, she thinks she might want to start reading.*

Quasim added this important observation:

> *There is really no pattern or formula to how one becomes a regular of the space. There is no process – one does not necessarily have to have the pre-existing love for books, a habit of reading or a knowledge or understanding of what the space is and its cultural, historical or political value.*

In the story Quasim tells, there is a movement from the every day, from habit and necessity (the student's responsibilities to her studies) towards an open engagement with reading, and thus with books and the bookstalls – that is, with the *boipara* itself. Her comments to Quasim about wanting to 'sit and spend time around in the area' suggest that she was already aware of the wider cultural currency that adheres to the *boipara* in relation to student life in general. Books and reading provide the aspect of the narrative that overtly points towards the possibility of change, of departure. She responds to an obvious disjunction in the connection she has made throughout her studies between the *boipara* and her reading and learning, *and* also responds to a slippage based in a more 'hidden' affective desire, a sense of becoming part of the atmosphere. Almost imperceptibly present behind the ready-to-hand formula that 'reading is a good thing to do' is this *desire* that actually provides the moment of slippage and rupture as well as the impetus that allows this young woman to want to undertake and understand reading in an entirely different way. In a conventional narrative, this might head thematically towards a binary tension between freedom and necessity. That binary *is* operating – indeed, it is precisely the extensive connection, the exclusive disjunction that *covers over* everything else that is going on for this student. When we look in-between freedom and necessity, we do not uncover a simple opposition or binary; rather, there is already a more complex movement at work towards a more nuanced change of her positionality from doing *what is expected of a student* (having the correct materials to allow one to prepare for classes and do well in exams) to *becoming a student* who hangs out in the *boipara*, and who therefore needs to read widely because that is what regulars of the *boipara* do.

In Quasim's narration, how she expresses freedom and choice in relation to books and reading initially has an air of uncertainty: '[S]he thinks she might want to start reading.' Nothing is decided, not even the choice to exercise her freedom to read. On the other hand, routine and obligation do not establish patterns that make their continuation or outcome predictable, as Quasim understands when his remarks move fluidly from the narrative about this young woman wanting to read books to the broader observation that, '*There is really no pattern or formula to how one becomes a regular of the space.*' *Both* necessity *and* freedom contain the potential for slippage and ruptures, and to see them merely as opposites cannot take us far. If we instead recognise the multiplicity of feelings, ideas and desires that actually inform the young woman's appearance at the bookstall in non-teaching time, even though she is not yet comfortable about articulating them herself, we also recognise how lines of flight generate, how identities shift, how trajectories change – that is, how ontologies interact with spatial movements and changing intensities.

What matters for the current discussion is that at the point where the material flow intersects with the body's pool of affects, *events occur*. I suggest that the student's emerging decision about her reading is a response to an intimation she senses of the haecceity of the *boipara*. The *thisness* of the place is recognisable in the books, not just as material objects but in how they carry ideas and education and the knowledge of 'worlds' of one sort or another. Clearly this is not the only measure of *thisness* here. It depends very much on the connections and relations that any group or individual carries into the *boipara* – thus *thisness* speaks to a very different apprehension for the booksellers, various workers, the radicals sitting around tables in the coffee house, the students who want to be regulars and radicals, the students who simply share their anxieties about exams or their futures around another coffee house table.

For the student who seems on a path towards reading for pleasure or at least for its own sake, it matters that she has begun to realise that books can reveal what they reveal, regardless of the intentions or status of whoever picks them up in the *boipara*. Confined by the image of herself as a responsible student (with all its attendant obligations), she seems to have caught sight of the possibility that books make their own readers and, further, that in the *boipara* readers become regulars. It is not what she might 'get out' of books (such as the possible answer to Question 12) that has now become most significant, but what they might *tell her*, a transaction that she recognises the *boipara* as enabling. For this student, it is a significant event to go to the *boipara* and to Quasim's bookstall for a new reason.

There are other occasions for slippage. Each night, when the booksellers pull up the shutters on their makeshift stalls, they need to move their books into a storage space from where, the next day, they will be brought back to be displayed again. There is nothing unpredictable about the routines the booksellers follow every evening. But the next day the books do not necessarily find their way back to the place or the pile where they were displayed the day before. As Rakesh told me:

> *Of course, we know what we have, or collected, or own – I mean, this is my business. However, you know, we do not arrange the books in exactly the same order every day. It is impossible to do that. We have old and new books, diaries, notebooks piled and packed and unpacked every day for business. On top of that, newer collections of second-hand material come to us every day too. People are coming every day to sell their old books to us. We sometimes reject the things that we have too many copies of and take in things that we know are priceless, or very valuable, and also at the same time those that are saleable, important for the students.*

One of the questions I asked all of the bookstall owners was:

> *When you have customers coming to your stall looking for books and other materials, how are you ever sure of what you have to offer?*
>
> *Rakesh*: *It is for this reason, we ask every customer, not only when they come into the stalls but also walk past the footpath, about what they are looking for. It is always easier for us to bring out the collections of all the materials we have if we know what the*

> *customers are looking for. Our stalls are really just stalls, and it is practically not possible for us to put everything on display. However, if you know what you are looking for and we know what that is, we bring it all out.*

There is another slippage, or a crack, in this – something that divides the orderly and regular procedures of bookselling from itself because of the particular nature of the *boipara* stalls. We know what book shops do, ordering their goods according to categories and genres of various sorts. Everything can be found by the customer as long as they know the code. The same is usually true of second-hand bookstores, although there may be different approaches to what should be categorised where. The *boipara* bookstalls, though, introduce something else through what we can call an intensification of circulation (as well as magazines and newspapers). Their materials circulate at a different pace; they circulate in great numbers compared with the size of the stalls (especially numbers measured by available floor and shelf space). And further, in their 'internal' circulation each night and each morning, books change places: they sort themselves into different piles; they are laid out in patterns that, from the customer's point of view, bear little or no relationship to where they might

Figure 6.3 Books spilling onto the footpaths

have been found the day before and where they might be found tomorrow. The continuous movement within and at the front of the tin walls makes the bookstall into a kind of haecceity, distinguishing the *thisness* of a space that is never stably itself. With his entirely individual cataloguing system, unknown to anyone else, only the magical abilities of the bookseller are adequate to the perplexing displacements of these books.

I can attest to those skills from my own experiences in looking for course texts in the *boipara*. There were times when I would arrive at a bookstall with not much more than the name of an author or a text. In the case of Milton's *Paradise Lost*, which I discovered in my second-year syllabus, I turned up at my 'regular' bookstall with little hope and a growing sense of terror about the consequences of not being able to find something so obscure. Suppressing the sinking feeling in my stomach, I asked, 'What do you have on Milton's *Paradise Lost?*' I was surprised by the profusion of what they identified as resources. The material they brought out for me to look at included Bengali translations of collections of Milton's poetry; summaries and ('made easy') study guides; books in English on Milton; exercise copies filled with notes made by other students on the works of Milton; and essays and reflective journals on *Paradise Lost* written by students who had studied the same course in previous years. That surprise (my affective response) was for me a small introductory window to what I now understand as the haecceity of the *boipara* bookstalls.

The neighbourhood and the circulation of knowledge

There are 64 colleges within the district of Kolkota, with three of those functioning under the title of the University of Calcutta found adjacent to or near the *boipara*. So it is easy to imagine accumulating over the years an extraordinary repository of lecture and other notes, essays and essay drafts, as well as scribbles on textbooks, numberless photocopies of research references and private tutorial materials, all carrying an equally extraordinary range of ideas, perceptions, interpretations and understandings dependent not just on variations in student abilities and interests but all sorts of other connections. If we move aside from the numbers and system for a moment, there is something significant to think about here. If all these students going to different colleges are following the same syllabus, they are all individually building their own notes from what is delivered to them as lectures in each of these colleges. Most students in most universities – especially outside India – probably end up storing their own contributions (even if these are now mainly accumulated electronically) for a few years, perhaps throwing them away eventually, or forgetting where they are to be discovered some time later, perhaps prompting a range of affective responses from nostalgia to embarrassment. But the *boipara* invites a very different choice for students in the vicinity. It is unlikely that every student of Calcutta University takes as many notes and collects as many associated documents as the most diligent do. But, as I have outlined earlier, a multiplicity of thoughts and ideas and references and cross-references – all sorts of ways of interpreting texts of no matter what genre that have been written down by students – arrive in the *boipara* to be passed on, to find themselves put to work and varied and understood or misunderstood by another generation, and another.

To put the consequences of this movement of words and thoughts another way, the *thisness* of student life – or at least of learning (since, just in the immediate context, the coffee house and its distractions still loom large), its haecceity – is recognisable in the disjunction between the general and the particular, that is the slippage between an apparently standardised syllabus and particular learning content. That might be found elsewhere. However, the unique haecceity of student life that the *boipara* leads us to identify is opened up by a slippage between the personal or individual and the shared aspects of the circulation of curriculum content and materials. University learning, which may take place in groups (the lecture theatre, the tutorial room) is still meant to be sedimented and assessed in the body of the individual; however, in the circulation of notes through the *boipara*, what the individual knows and understands becomes a group possession, passing from hand to hand. It does so across colleges and across time. So, the other aspect of this is that the circulation of such material through the *boipara* unsettles the assumption that it relates only to the context in which it is collected (the course the student is enrolled in and how it unfolds in that year or semester). In the *boipara*, all course notes and all research materials might concern themselves with all possible variations in the teaching of a syllabus.

Like any university district, the *boipara* had – and has – its share of eccentrics. Krishna Bandyopadhyay (2014, p. 168) recalls one:

> There was a character called Chacha at the coffee house. He could not imagine that anyone would not be able to enter the place. If it was full and a group was leaving because there was no place free, he would run up to them with a chair and tell others to leave or move around so that everyone could participate. He thought himself as a kind of guardian of the place and he had such a warm and open laugh. One day, he came to know that the police were coming to arrest someone. To him, the politics was not important, but he could not bear the thought that any coffee house regular was to be arrested. So he came and warned the person.

As I have mentioned, part of the 'back story' to this episode is that many students and staff of the university and other educational institutions close to the *boipara* were members and sometimes key leaders of radical political groups at various times. As a result, in the 1970s, for instance, the government as well as the police kept a very close eye on people entering and spending time in the *boipara*, in particular in the university canteens and the coffee house. Interestingly, Chacha maintained his own surveillance of the clientele of the coffee house, unauthorised by the proprietors or by the government but nonetheless a kind of individualised mirror of the state panopticon. Nothing in the comings and goings at the coffee house seemed to escape his attention (which is ordinarily the burden, of course, of the security services and, more innocently, of the owners of cafes).

Both the owners of cafes and the police are bearers of 'authority', even though the authority of the police might be more daunting. Both can interrupt the 'natural' order of the coffee house established by the habits of the customers. This is true of coffee houses in general (according to which people come and go, tables are sometimes full, people sometimes stay too long, others leave and go elsewhere if there are no seats or tables). In the ordinary course of events, neither form of authority needs to make itself particularly visible in a situation like the coffee house. Only when things get out of order is it necessary for a proprietor to assert control. The police, on the other hand, do not usually need to be seen in such places. As representatives of state power, their authority is taken for granted. Chacha, though, was a free agent who mobilised the procedures of authority and surveillance belonging to the proprietors and to the police. He also subverted, in Bandyopadhyay's description of his behaviours, the proprietors' responsibilities to maintain orderly premises by substituting an order of his own. And he subverted the reach of the law expressed in the reported intentions of the police to arrest someone.

We might be able to call Chacha a singularity – neither proprietor nor police – but, because he appropriates and circulates between those positions, he is also a sign of slippage or disjunction between them. In this, Chacha occupies a trajectory similar to that of the book in the example of the *thisness* of the bookstalls in the *boipara*. That is, he produces a connection between coffee house and police through his adaptation of the different categories of surveillance and occupies the zone between them. But in his subversions of both categories, he also produces a slippage or crack in the connection itself. The coffee house is carried outside the regularities (commercial and social) of surveillance, and in its movement we can catch another glimpse of a *thisness*: the haecceity of the coffee house is attached to a recognition of it as a site 'outside the law' – a site, static though it may be materially, where ideas uncomfortable or threatening to the state and its good order, or plans for its subversion or radical reform might (as they have in the past) flow back and forth across tables, and from one table to another and finally out the door to circulate in the city and from there in the nation beyond. We might also want to ask: what is the *boipara*'s part in all this? Since everything seems more or less focused on the coffee house, the *boipara* might easily be taken as no more than a location, the background to what is being played out. But I have already mentioned how the police in Chacha's day also kept the *boipara* under surveillance, so College Street was obviously far from 'innocent' as a venue for radicals and subversives. Nor was it free of dangerous ideas. As we have seen, there were (and are) the books and the magazines and the papers, vast collections with all sorts of ideas in them, including dangerous ideas. The haecceity of the coffee house, then, reveals itself in between the politics of what has just happened and the politics of what is to come (since ideas have a past but are always potentially disruptive, even when shut up in books that constantly change their places in the stall, between stall and customer, and between stalls when customers bring them back to different ones).

Writing in the space while being in the space

Sitting in the coffee house today, as I begin to write about the boipara, *I am reminded of an anecdote. This was during the early days of our university lives when we made a trip here with our college seniors. Rashmi di, a brilliant scholar, painter, sculptor and our immediate senior at the department of English in Scottish Church College, virtually took it upon herself to give us a 'brief' about what it was like to be a student of literature and come to the space. I remember her saying, 'You have to give time to the* boipara *for it to make you one of its own; and then slowly, gradually you will keep coming to this space; you will acquire the skill to be able to think, to write in this space while you were here amid all the cacophony.' At that time, this surprised me. Why was she so insistent on the fact in order to become one of the space, experience the space, writing within the* boipara *was important? Also, talking about the noisiness of the* boipara, *she insisted that this collective soundscape formed from the* adda, *the sound of the traffic outside, the chatter of the bargaining going on in the stalls outside would all quite suddenly become familiar, known and almost enjoyable. I had heard many stories of how eminent Bengali poets like Sunil, Shakti and Sandipan had spent hours in the coffee house, not just indulging in* adda *but also using the space for their creative inspiration. There are tales of famous Bengali singers composing tunes, scribbling couplets on the back of the receipts of their bills for the day at coffee house. Sometimes, they would leave behind these papers forgetfully and the waiters would later find them and store them and give them back on their next visit. In no time, this became known to everyone in the coffee house. As a result, young poets would often wait until the more prominent writers left and, once they did, would often go and look for their leftover notes in receipts, or other bits of papers in the table.*

It is hard to track the historical 'authenticity' of these stories and encounters, but somehow, in coming to one through a number of sources who pride themselves as insiders to the space, they seem believable. Or, I ask myself, do we believe in these stories because we want to? What is it about this space that makes us want to believe in these stories and yearn for more? For example, there was this one time when me and my friends were discussing whether or not the famous Bengali film maker Satyajit Ray[5] *was indeed a regular of the* boipara *and the coffee house. One of our friends, Deepika, recounted how she had heard from her mother how she had indeed seen Ray in the coffee house with his friends. In the meantime, Moni da, the waiter who was serving us coffee, arrives at the table with our coffees and* pakodas. *He overhears our excited conjectures about whether or not Ray was indeed a regular. He pauses and adds into our discussion,*

> *I have been working here for as long as 30 years and yes, I have seen him spend time here. Sometimes he would bring his sketch book and work on his sketches by himself. Other times he had company. Like you we have had many others wonder whether he did his sketches of Pather Panchali or Feluda sitting in the coffee house. But you know it is not just filmmaker Ray or singer Manna De.*[6] *Scientist Satyen Bose*[7] *was also a regular of the coffee house and many often wonder whether he had drafted his communication with Einstein for the Bose–Einstein condensate here, or whether Nobel Prize-winning economist Amartya Sen, who was also a student of Presidency College, has had some of his ripe economic discussions accompanied by the coffee and chicken Afghani of the coffee house.*

Moni da then walked away to other tables, taking more orders and joining in on their adda *topics. This is the thing about the* boipara. *Today, as I sit in the coffee house to write about the* boipara, *all of these stories and experiences become important and relevant to me. I wonder if these have effects on my writing.*

There are a few strands of thought on which I want to reflect from these field notes. It is evident, through the stories that account for the creative work that artists, scientists and filmmakers such as Ray, Bose or De have produced while being present in this space, that the *boipara*, through its storage and circulation of knowledge, generates an atmosphere where creative activities flourish. Therefore, the space has a material, intellectual and affective impact on the people who use it, which facilitates these occurrences. Several times in my college days, I witnessed my friends specifically come to College Street in order to write. The spatial environment of the *boipara* and the coffee house triggers a moment of *making sense*. One is inspired not just to experience and internalise the space but also to be or become creative in it. There is something along the lines of a kind of spilled energy that provokes creative expression, be it writing poetry or songs, or making sketches or even simply indulging in productive discussions. All these activities become eventful. During my field trip, on the days I decided to write about the *boipara* while being there, my thought processes were also filled with everything that I had heard about the process of writing, creating while being in the *boipara*. This overwhelming *presence of the space* influences, guides and navigates our capacities to think *about* the space. This is often reflected in how we write about the space.

What I am trying to describe is something that felt to me like an actual, animated conversation between what I was seeing unfolding in front of me and what I had heard about the space. This is not to say that the creative work of the likes of famous artists like Ray or Gangopadhyay created a burden – like a pressure to produce work of that standard. The role of memory and affect did not pressure us to produce brilliant work, but just to produce work – or in my own case to keep writing. It is why, when designing my fieldwork, I purposefully set aside days when I would only sit in the coffee house or tea stall and write about the space *as I see, feel and experience it*.

The students' rally today is scheduled to leave the Presidency University precinct around 11 am. The boipara *feels a bit more politically charged today than on the other days I have been here on fieldwork so far. People are anticipating that all sorts of things will happen. The traffic is already getting busy; the gradual increase of the students' chatter is adding to the overall soundscape of the space. I am about to go and take a walk along one of the interior laneways of the* boipara *but I make a note to myself to come back in time to witness the students' protest walk.*

After a few minutes of walking in the laneways, I come back to watch the student protest. The protest is about the Central Board of Film Certification unnecessarily blocking the release of certain films that do not endorse the political views of the ruling government. The students, many of whom are also amateur filmmakers, are passionate about this. Watching them walking along the main road, blocking the traffic, chanting slogans passionately, was too tempting. I have done this so many times before as a student. Without thinking any further, I join the march and walk along with them. I come back to the tea stall later. I am contemplating writing something about it while I am here. As I am also doing fieldwork, I ask myself, why is it that I feel that I must write here? Why not go back home and write about it some other day? I realise with more force than ever before that the act of writing about the boipara *while being in the* boipara *has a different effect on the process of writing. It is the coming together of all the parts and participants of the space that prompts us to a realm where the words that best describe or rather translate our affective experiences, visual experiences, audible experiences are closest to what we have been feeling. In this manner, writing becomes eventful.*

Thus:

After many years I feel at home here among the dusty, smoggy, dirty yellow hues of the street. I see my juniors walking on the hot, blistery concrete tram lines of the main road of the boipara *and it gives me immense pride. I stop to wonder why I am feeling a sense of pride. I realise that it is because the first-hand, second-hand and indeed many-handedness of the books, the depth and power of knowledge that has been carried in this space for decades, always finds ways to mobilise itself through the students, scholars, and owners of the book shops and bookstalls of this space. The people who buy and read books from here internalise the knowledge they gain from being in the space, forming not only a personal, intimate relationship with the material components of the space but also an emotional allegiance. I cannot help but wonder whether this is why I have come back to the* boipara *as well. The childhood memories of occasionally walking through these very streets with Dadu and Baba fill my senses. It was this place that gave me one of my earliest childhood thrills and adventures; it was this place that awakened my first political self. I realise there is a sense of gratitude and affection that we all feel towards this space. This nexus between circulation of knowledge and scholarly practice of the knowledge has a distinct effect on us. I had only planned to sit and watch the protest march, but being in the space I could not help but join in too. This yearning to be a part of the space, by taking part in the protest march, writing about that experience while in the space, is triggered by a sense of mobility that is eventful. When in the* boipara*, we become an active part of it, doing the* boipara *as it unfolds. The affects that we experience prevent us from being still, passive observers; they inspire us to do the space – through walking, browsing books, taking part in its political processes and even through writing about it.*

What is it to write about space? Is it different from writing about anything else – objects, emotions, events? I have dwelt on the issue of writing more than once already, but because this writing is part of my engagement with the *boipara*, I continue to reflect on how I am developing something that is not just a report on (i.e. representation of) those engagements. Returning to my own process of writing has also helped me to reflect about how writing has been and is very actively engaged in the mobile processes of the *boipara* as an assemblage and how it connects to assemblages beyond itself. Although writing is conventionally regarded as a representational medium, I have come to recognise how the unfolding of space is best explained through trajectories of thinking that are different from conventional forms of representation and conceptualisation. There is clearly a potential roadblock when efforts to escape the conventions of representation need to be expressed according to what seem to be the terms of those conventions. Perhaps we need to be creative in asking questions around writing about city spaces such as the *boipara*.

Exploring a space that is primarily characterised by movement, shifts and relationality requires that we try to find new or renewed approaches to writing/writing about everyday spatial experiences. I do not think that it is productive to question the utility of scholarly writing, nor is it useful to argue that we should replace it with 'creative writing' as something assumed to be its binary opposite. If there is anything we can learn from wandering among the bookstalls and browsing in the many forms of writing in the *boipara*, it is surely that such a binary is not only problematised, but rendered meaningless in the context. Even though Dadu was a particularly well-read man who held scholarly activities in very high regard, his modes of valuing of writing operated on a markedly continuous plane and through heterogeneous strata. I am suggesting that we start to think in terms of experimental approaches to scholarly writing that are entirely comfortable weaving together personal experience, affective knowledge and conceptual reflections to make multiplicities and open new conversations. After all, the processes of making multiplicities are a significant part of why those of us who have great affection for the *boipara* also have great affection for second-hand books and for the many-handed evidence of other people's thoughts and feelings that they carry in the writing in their margins, on flyleafs, between chapters and paragraphs, on slips of paper, bookmarks, accidental backs of shopping lists and so on. These become imaginative links and departure points for everyone engaged in the circuits of knowledge to which we all contribute. They participate in the business of making sense in myriad ways.

Recent decades have seen a range of approaches to thinking and writing that destabilise or set in motion fixed and static representations of space, notably by Soja (1996) and Massey (2005). Without attempting to cover the whole terrain that opened out from the 'spatial turn', it was bound to impel such further developments as: the 'material turn' through actor network theory (Latour, 2005), thing theory (Brown, 2001) and new materialism (Bennett, 2010; Coole & Frost, 2010); the 'new mobilities paradigm' (Sheller, 2014; Urry, 2016); and the development of 'non-representational theory' (Thrift, 2007) for thinking about spatiality. All of these have contributed significantly towards thinking with and through different forms and systems of representing space. No single theoretical approach has settled the matter, however, and the problem of representation remains open in many different disciplines, including cultural geography. Ben Anderson (2018) discusses the idea of the 'force of representations' in the context of renewed interest in the question of representation in cultural geography.

Whether in relation to how new genres of climate art might spark response to anthropogenic climate change (Hawkins et al. 2015), the role of digital images in the ongoing (re/de)composition of urban life (Rose, 2016), or the functions of talk and text in 'fixes' to mobility infrastructure crises (Bissell & Fuller, 2017), there is a concerted effort to understand the force of representations as they make, remake and unmake worlds (Anderson, 2018, p. 1). Various returns to the question of representation in relation to writing about space continue to open new creative lines of thinking. This shift is a response to developments not only in language theory but also in our understandings of space and spatial practices, which require us to frame our interactions with places in diverse new ways.

In the light of all this, navigating the space of the *boipara* (a series of experiences that I like to refer to as 'doing' the space) and writing from those experiences (in ways that hope to escape representationalism) are not easily separated. In this, I find myself echoing Anderson's (2018, p. 3) observation about the importance of considering 'representations (in all their diverse forms) as only ever part of and becoming with a host of other processes, events and things'. From this perspective, my experience of the *boipara* is part of my writing and my writing is part of my experience, so it cannot be produced as a record of that experience or regarded as such. In writing my own fieldwork observations and reflections, I consciously tried to break with my own long-term habits of scholarly writing, instead allowing myself to use words and phrases that expressed feelings and sensations. As often as I could, I wrote my notes in the midst of the events unfolding around me; however, when I wrote reflective diary entries later, I still tried not to slip into 'scholar' mode, marked as it is (in the way I was trained) with a completely unrealistic effort to *be* 'objective' by *sounding* 'objective'. My conscious awareness of expressing feelings, affective responses, emotional memories and so on frequently brought my reflections close enough to what was happening to allow me to recognise my interactions with my interviewees, and their interactions with each other, as forms of expression *in themselves*.

Notes

1 As explained earlier, *didi* is a term of endearment often used in public spaces to call to/draw attention to someone who identifies as a female (including female customers and female researchers).
2 In his book on Leibniz, Deleuze (1992) considers at length the philosophical question of 'the fold', but I am applying 'fold' in the everyday sense and, usually, as a verb or participle (to fold, folding and unfolding).
3 Although, in the distinction drawn by Deleuze and Guattari, we find some subjective or personal recognition of this oscillation in *affections*, the particular ways in which individual bodies may be changed by what happens.
4 Question banks are collections of questions that have appeared in central or state board examinations of previous years. These are similar to workbooks that have compiled lists of questions that appeared in Grade 10 and 12 school exams, undergraduate and even postgraduate exams, from the last five years. They are very popular among students, as these books give them an idea of what to expect when they are preparing.
5 Satyajit Ray is one of the biggest and most successful filmmakers in the history of Indian cinema. His most notable work, *Pather Panchali*, has won numerous international awards and is still studied and researched along with his other works. He is the only Indian filmmaker to have won an honorary Oscar for his contribution to cinema. Apart from being a prolific filmmaker he was also a musician, writer and sketch artist.
6 Manna De was a famous Indian singer, composer and writer who has a significant body of work in both Bengali and Hindi music. Here, I refer to his famous song '*coffee house er shei adda ta a jar nei*'. He was also known to be a regular at the *boipara* and coffee house.
7 Scientists Satyen Bose and Jagadish Chanda Bose were eminent physicists who were also regulars of the space. Satyen Bose is best known for his contribution to early quantum physics that garnered huge international attention. His collaboration with Albert Einstein, famously known as the Bose–Einstein condensate, is being referred to here in terms of his communication with Einstein.

References

Anderson, B. (2018). Cultural geography II: The force of representations'. *Progress in Human Geography*, *43*(3), 1120–32. https://doi.org/10.1177/0309132518761431

Bandyopadhyay, R. (2014, 30 May). The hawkers' question in postcolonial Calcutta. *Modern Asian Studies*, *50*(2), 675–717. https://doi.org/10.1017/S0026749X1400064X

Bennett, J. (2010). *Vibrant matter: A political ecology of things*. Duke University Press.

Bissell, D., & Fuller, G. (2017). Material politics of images: Visualising future transport infrastructures. *Environment and Planning A*, *49*(11), 2477–96. https://doi.org/10.1177/0308518X17727538

Bonta, M., & Protevi, J. (2004). *Deleuze and geophilosophy: A guide and glossary*. Edinburgh University Press.

Brown, B. (2001). Thing theory. *Critical Inquiry*, *28*(1), 1–24. https://doi.org/10.1086/449030

Colebrook, C. (2002). *Gilles Deleuze*. Routledge.

Coole, D., & Frost, S. (2010). *New materialisms: Ontology, agency, and politics*. Duke University Press.

Deleuze, G. (1990). *The logic of sense*. Trans. M. Lester & C. Stivale, ed. C. Boundas. Athlone Press.

Deleuze, G. (1992). *The fold: Leibniz and the Baroque*. University of Minnesota Press.

Deleuze, G., & Guattari, F. (1987). *A thousand plateaus: Capitalism and schizophrenia*. University of Minnesota Press.

Hawkins, H., Marston, S., Straughan, E., & Ingram, M. (2015). Arts of socio-ecological transformation. *Annals of the Association of American Geographers*, *105*(2), 331–41. https://doi.org/10.1080/00045608.2014.988103

Latour, B. (2005). *Re-assembling the social: An introduction to actor network theory*. Oxford University Press.

Massey, D. (2005). *For space*. Sage.

Rose, M. (2016). A place for other stories: Authorship and evidence in experimental times. *Geohumanities*, *2*(1), 132–48. https://doi.org/10.1080/2373566X.2016.1157031

Shaviro, S. (2007). Deleuze's encounter with Whitehead. Retrieved from http://www.shaviro.com/Othertexts/DeleuzeWhitehead.pdf

Sheller, M. (2014). The new mobilities paradigm for a live sociology. *Current Sociology*, *62*(6). https://doi.org/10.1177/0011392114533211

Soja, E. W. (1996). *Thirdspace: Journeys to Los Angeles and other real-and-imagined places*. Blackwell.

Thrift, N. (2007). *Non-representational theory: Space, politics, affect*. Routledge.

Urry, J. (2016). *Mobilities*. Polity Press.

7 Conclusion

Introduction

I would have preferred to say somewhere (perhaps here) that this book has no beginning and no end, a claim also made by Deleuze and Guattari (1987) in their introduction to *A Thousand Plateaus*. But that might easily have been taken as some kind of textual imitation of their work, which is not really in the spirit of Deleuzian thinking. I might also need to mention, since I have been writing about an urban space – a practice that inevitably falls within the shadow of Michel de Certeau's (1984) pivotal work 'Walking in the City' in *The Practice of Everyday Life* – that I do not have such an accessible convergence available between writing and my own journeys in the *boipara*. I am simply too familiar with College Street. There is, for me, no getting lost there and no straightforward way to frame a representational response to my experiences. If I had wanted to take my cue from de Certeau, there would also be a problem in dealing with the books. If it might have been possible to chart a similar kind of interrupted and hesitant walking in the *boipara*, one of the influences on changing direction (frequently, constantly even) is the piles of books, which present as a series of material obstacles to movement through the space. That might be easy enough to register physically, but there is – as I hope I have shown – much more to the books than their bulk in terms of pages, because their content, their ideas, get in the way in another fashion and clamour for recognition from the passers-by. At any given moment in time, the books open pathways to divergent imaginations, ideas and conjectures. At the same time, the booksellers are constantly producing new pathways in knowledge for students and other regulars as they are circulated with all their notes and marginal comments.

The word that comes to mind most often in trying to characterise my experiences of the *boipara* is 'wandering'. Yes, it has always been possible – sometimes even necessary (especially as a student) – to go there with purpose looking for a particular book or other resources. But even then, it was easy to pay attention to distractions, to sit and read books I was not looking for, to go to places I was not intending to visit. This is something that I think is useful to apply in describing my method in this book. Not that I have tried to produce some form of textual metaphor for or imitation of my own wandering (which,

DOI: 10.4324/9781003293026-7

again, would be with de Certeau as a reference point); rather, I have attempted to develop a form of writing that draws attention to the interactions, the cross-overs to which I have just pointed by identifying the reading of books as part of the processes of that wandering. I have not attempted to provide any kind of systematic account of Deleuze and Guattari's (1987) theoretical perspectives or to 'apply' these in a methodical or complete way (this, too, would be against the spirit of their work). Rather, I have developed a wandering with and through their works – again not in any metaphorical or imitative fashion as a mirror to my physical wanderings, but picking up issues and concepts as a response to moments in my experiences (or those of others) of the *boipara*. So what we have is a wandering in the ideas of Deleuze and Guattari, not taking place alongside and as a reflection of the wandering in the real world of the *boipara*, but taking its momentum from crossing back and forth between that world and the world of Deleuze and Guattari's writings (which are just as real), as well as the writings of a number of others.

In this context, I need to recognise again the importance of the idea of the rhizome to my work, to my wanderings. Rhizomes emerge from the milieu, the in-between, the middle. Those who understand rhizomic movements understand that everything begins in the middle. My familiarity with the *boipara* means all my wanderings there and all my efforts to account for them rely on a sense of stepping into the middle, finding myself on trajectories that emerge in the middle, and noticing connections that generate from the middle. As I mentioned in the introduction to this book, my Dadu's and Baba's toing and froing with books for me and for themselves meant my engagements with the *boipara* have always already begun. But the uniqueness of the *boipara* is that its particular physical nature insists that to step into the space is to step into the middle. Everything is in a sense already related to everything else, whatever the layout suggests, which makes wandering what it is – often with movement in unchosen directions. And so, even without searching for ways to represent it as such, the rhizome is a continuing reminder that there is no beginning or end to my engagement with the *boipara* (with *any* engagement with the *boipara*).

To use a word such as 'wandering' might suggest that it is expected to work here as a trope, another participant in the circuit of representation. In turn, that would imply a relationship between the physical world of the *boipara* and a domain of thought and ideas that flows only in one direction. But I hope it suggests much more than that, since there is an equally significant question about what wanders from thought and ideas towards – indeed into – the *boipara*. This is not a question about whether such movement might lead to a 'deeper' theoretical exploration (which might well be possible in a different kind of book), but rather about how ideas and thoughts join themselves to the *boipara*, make it part of how they recognise the world. In other words, how does wandering help us to recognise that the *boipara* is not some adjunct to knowledge and thought, serving the interests of the university colleges that hem it in, but a participant in the processes of knowing and understanding.

Identifying the importance of difference is crucial too. In relation to thinking about a space such as the *boipara*, it is necessary to recognise that our processes of spatialisation (how we conceptualise the space) rely on understanding the role of both difference and repetition, terms whose relations and interactions Deleuze (e.g. 1995) in particular made familiar and from which I have drawn from time to time during this wandering. Every time we think about or revisit our experiences, we repeat/relive those moments, but this is a repetitive process in which *something is always different*. Something always changes about the known and repeated routes, memories, feelings, insights, material encounters, embodied interactions. There is always a deviation or a detour; always a transformation. Sometimes these are small and subtle changes, induced by imagination or thinking itself. Sometimes they are a response to a sense of in-betweenness: an emerging awareness of a gap or a moment of not grasping or understanding opened up when we tally our known experience with the unknown and are forced to speculate or guess or anticipate.

There are numerous ways of thinking, experiencing, understanding and writing about city spaces such as the *boipara*. Each of these ways, depending on the discipline from which its frameworks are borrowed, comes with its own style, structure, analysis, means of validation and limitations. In the process of deciding what approach worked best for me to develop/express my understanding of the *boipara*, I found that aligning myself with one or two or more disciplines quickly began to limit what it was possible to do because of the problem of frameworks themselves, and the fact that these could not expand sufficiently to contain or account for my lived experiences, for the affective resonances and sensorial registers involved in my encounters with the space and its material and embodied occupants. It is for this reason that my writing follows – and perhaps on other occasions also leads – the ways in which my experiences have made sense to me. My chapters have been developed as an expression (and, as little as I could manage, as a representation) of my own wanderings in the *boipara*. Of course, that means the book is only one of the many possible ways in which I have experienced and will continue to experience the *boipara*. It is in this sense that, whether referring to Deleuze and Guattari or not, I need to go on claiming that it has neither beginning nor end. In my wandering – a wandering alert to the inevitable intensification of the assemblage of the *boipara* – it is in my fragmentary encounters – in bits and pieces – that I have been able to enact and re-enact, live and relive, imagine and speculate about what it is to be a part of this space and what it is to think about it as well.

The spatiality of the *boipara*, as I have attempted to capture it, is grounded in a unique movement – a circulatory trajectory that is characterised by reuse, renewal and emotional attachment. Through and within the living stories memories and attachments with the books, the bookstalls and other characters within the space, I have demonstrated the ways in which the heritage of this space, and heritage in general, is not something that is stagnated in the past. It is mobile, affective and powerful in its influence in the present and its potential in the future. The practice of reuse and renewal is often relegated to necessity (Gregson & Crewe, 2003) and thrift shopping and other practical social reasons. As a result, the emotional and affective value of these encounters become fleeting and is missed. Their value over time and their importance in thinking of the heritage of the space are overlooked.

These encounters have also begun, paused, stopped, restarted and continued in no fixed order and according to no particular expectations or projections. That is, however, not an abandoning of what might be called a research methodology since this has been contained in the noticing and the analysis – and the various crossings into and out of theory – provoked by whatever has taken place. If these, too, remain incomplete, the process of writing has provided a comforting realisation that it is alright – in fact, it is useful – to be inadequate in apprehending, interpreting and representing stories and experiences. I hope this book is a result of listening closely to and thinking honestly about all I have experienced in my engagements with the *boipara* and with those whose experiences have taught me so much about the place and what it means to them. And if whatever map I have made of their experiences and my own is, like everything else, unfinished, I also hope its multiple and open-ended pathways will continue to lead elsewhere, in thinking, in spatialisation, in how we write about and create knowledge. If so, then I will have done something that does not represent or portray the *boipara*, but instead responds to it.

Figure 7.1 The *boipara* continues to unfold

These multiple and open-ended entryways show potential for further thinking, for spatialisation, methodology, writing and hence knowledge creation. I hope the process continues.

Earlier, I used a diary note about walking to the *boipara* with Dadu, his enjoyment of walking through the back lanes where the publishing houses and printing presses are, and his obvious affection for and fascination with all the processes involved in printing – ink, presses, paper and so on. The diary entry below is part of the rhizomics of my family and their connections to an assemblage that might be seen to 'explain' much about Dadu, but that actually provides another in a multiplicity of perspectives (most of which are absent from this book). It offers another entry into the middle of bits of space–time and another recognition of potential departures, lines of flight. I believe that spaces like that of the *boipara*, being true to their essence, have no beginning or end. Thus, value of this space as a heritage site is never stagnant – it is creative and in motion. The *boipara* is never only a historical place stagnant in time. It is living in its practice where heritage is not only felt and experienced but also reimagined. I end this book with another excerpt from my fieldwork experience. In doing so, I leave open a door that invites the readers to immerse themselves into this space.

Reflective diary notes

After a week's break, I am back to the boipara. *The walk to this place is never the same even though one always knows that the* boipara *never fails to give the experience that one anticipates. It is a feeling of knowing and not knowing at the same time. I decide to walk to the* boipara *today taking the route that Dadu used to take when we went to the place together. Wandering on the footpath alongside the tram lines, still about ten minutes away from the Bidhan Sarani crossing, I wonder, why did Dadu have such an intimate and affective relationship with the space? Could it have been our printing ink supply business? My great-grandfather in the later nineteenth century had established a well-known printers' ink supply business in Calcutta's China Bazaar. Later on, Dadu had taken over from him after my great-grandfather's demise. Among other things, I particularly remember the ways in which Dadu used to talk about the ink and its association with printing. It was as if his relationship with the printing ink never ended with it merely being a material commodity with which our family did business. He always talked about the processes with which the ink used to be transported to the printing presses; the publishing houses to which our business supplied the ink were also important to him. I remember we used to get free copies of books and calendars from all of these printing houses every year. Dadu stored all of these carefully in his study. In his walks along the* daftaripara *that is close to the* boipara, *he often visited the printing press houses where our business used to supply ink. I also remember noticing how, every time he got hold of a book, he commented on the quality of the paper, the ink, the typesetting among other technical details. I often wondered how his relationship with the* boipara *was materially so detailed and minutely etched out.*

While Baba's relationship and even my relationship with the boipara *often oscillated between our lived experiences and what we absorbed from the content of the books, Dadu's relationship was slightly different. His close association with the* boipara *was intricately braided between the materiality of the books, his own occupation as an ink supplier to those who printed the books and other materials that got printed in the* daftaripara*; later, the circulation and movement of those books – first as new copies in the book stores and later as second-hand copies in the book stalls – all of these layers of connections conversed with each other, for Dadu, at the same time.*

There was also another reason why Dadu had a special place, emotionally, for the printing press houses near the boipara *and also the stalls within the* boipara *that exclusively stored Bengali books. Dadu had an affinity towards reading Bengali literature over English. This had very strong influence on not only his process of collecting second-hand books but also his walks with me in my early days in the* boipara. *Carefully exploring the boipara, he used to always take me to the stalls that had only Bengali books. Actually, the matter of what kind of books I would buy from the* boipara *was a subject of debate between Dadu and Ma. Ma, who herself had an honours master's degree in Bengali literature from Calcutta University, was not very keen on me reading Bengali books. She was very happy about her own achievements but had a very different view about what my education would entail. She often ensured that I read Enid Blyton's* Famous Five *and* Secret Seven *books as a child over Sukumar Ray and Abanindranath Tagore, among others. There was no aversion to Bengali literature but definitely a colonial fascination with its English counterpart. Dadu, on the other hand became increasingly concerned about this situation. He was often worried that I was not reading enough Bengali. Thus, he often took me, when I was a bit older, to the stalls that had Bengali books. When I used to bring back a collection of Bengali books, Ma used to smile; she was still happy that, like everyone else in the family, I was getting used to reading books, but secretly she preferred that I read more English than Bengali.*

As I reach the boipara, *I bump into one of the workers from the Durga printing press: Mallick Babu, with his signature ear-to-ear smile never fails to recognise me even after this many years. Exchanging pleasantries, he complains,* 'tomra to ar ashoi na' *(you people hardly ever come back to the* boipara *and our printing press house, once you leave the city).*[1] *I sense a combination of love and sadness in his voice. His expressions indicate that the agreement between users, regulars, customers of the space and the booksellers, to come repeatedly, pay a visit, spend some time in the* boipara, *is not a one-way emotional assumption that we make. It is an expectation and a desire that is very much preserved and expressed by the booksellers, the printing press house people and other current regulars who have known me and my friends and family for a long time. Somehow, Mallick babu's complaint came wrapped in a sense of warmth and the feeling was comforting. Slowly walking along the footpath of the* boipara, *I bumped into Sukumar, one of my dear university friends. Sukumar, has not changed at all: same old scruffy hair, khadi Panjabi, torn jeans and a rug bag as his primary companion. We hug each other and before I can begin a conversation, he stops me abruptly and says, 'How long are you here for? Are you free now? We are meeting at the coffee house for an* adda. *It will all be familiar faces, me, Aditya, Reshmi, Saheli, Payel, Ranit da, Dev da. Save the pleasantries for then, you bastard.' Sukumar invites me to the* adda *with his signature smile and slang.*

During my fieldwork, I had made multiple visits to the coffee house, spending time sometimes by myself and sometimes with the friends with whom I was able to get in touch. However, it was never with this group of friends that I am about to meet today. They used to be my core group in university days. We had our first drinks, cigarettes, political rallies, crushes, breakups, exams, fights and graduations together. Coincidentally, a lot of those significant life events occurred in the dingy lanes and by-lanes of College Street. Quite naturally, the sense of anticipation and excitement is on another level. Upon climbing the winding staircase of the coffee house, I bump into them. I am welcomed by warm hugs and the choicest of slang, some coated with sarcasm, with love. Finding ourselves a table and ordering the first of many rounds of our customary 'kalo cha' *(like most other people in the coffee house, we were black tea or* 'kalo cha' *loyalists), we make ourselves comfortable. Reshmi and Saheli, like me, are in academia and are in the finishing stages of their doctoral degrees in the United States. It is because of this career choice we are treated with enormous amounts of mock sarcasm and disdain in our group – we were the 'sellouts', according to Sukumar and Dev da. The sarcasm still has not changed. Sukumar and Dev da are still regulars of the* boipara *and the coffee house. Jokingly, Dev da mentions how the nature of students' politics was fast changing on the college and university campuses and this was having an effect on the* boipara. *The left was gradually being replaced by the conservative right in the last few years. Fewer and fewer people were standing up for the students, the bookstall owners and other stalls in the space. Any form of art or creative expression that dared to speak out against the central government was being labelled as anti-national, unpatriotic. Dev da has a small crew of filmmakers now. He writes scripts for short films. As a continuing regular of the* boipara, *he organises street plays and screens his films in the lanes and by-lanes of the* boipara, *and even in the canteens of the universities in the College Street precinct. He told the story of how he was refused a screening at the university campus of his documentary on the radicalisation of Hindu nationalists through university education. As a sign of protest, with the help of some bookstall owners and students from the university, he was able to have an alternative screening on the streets of the* boipara.

There is something about the ways Dev da speaks that makes us – myself, Reshmi and Saheli – feel a bit guilty about leaving the boipara. *But we hadn't really left the space; it still had a very strong relationship with us, in our own ways and on our own terms. Sukumar is a budding Bengali poet now and also a professor of Bengali literature at Calcutta University. I could hear in his tones, too, that they took pride in staying back, cultivating what we all used to do in the space once upon a time as professions and most importantly being active and still a regular in the* boipara. *Our conversations move forward, swaying from animated discussions on politics to the food in the coffee house. Every now and then, we see familiar faces walk in and out of the coffee house. About an hour and a half later, we take the* adda *to the tea stall outside the coffee house. I remember it was this very tea stall where Saheli introduced me to the poetry of Rainer Maria Rilke. It was an old, translated, second-hand version. Saheli had just had her first serious relationship breakup and she had taken up the bohemian, rich, imagery-based poetries of Rilke to heal her. I remember vividly that it was a grey afternoon, where the heavy rains were providing some much-needed relief to the heated tram lines of the* boipara. *I was seated inside the coffee house when Saheli came rushing with a copy of the book inside the stall. We read 'Before Summer Rain' and other poems all afternoon, kept flipping through the book, drank hot tea, enjoyed the rain, looked at other people taking shade in other stalls, watched the trams pass by, people rushing to get off and on the tram with their half-opened umbrellas.*

Saheli sat next to me in the teastall today, she spoke about life in the United States and I spoke about Australia. We shared our experiences of being postgraduate students and life in general. She was excited to know about my writings on the boipara. *As a passing comment, she mentioned that she still had the book of Rilke's poems. We did not revisit that afternoon; however, we spent a few minutes of silence. Later, she suggested that we should write something together in the* boipara *the next time we are here. This moment was important for both of us. It was reminiscent of the numerous afternoons we all had spent, together and individually, in the* boipara, *of how each book we picked up from the bookstalls, read in the tea stalls or coffee house or even at home shaped us and redefined our relationship with the* boipara. *However, most importantly, it is these memories of spending time, reading books together in the space that remains etched in our memories. They become active and create newer connections every time we are back in the* boipara.

Our adda *lasted until about eight in the evening. I reluctantly indicate that I have to head home. One by one, all of us left. I decide to take the tram home. While at the tram stop, I noticed that some stalls were still packing up their second-hand books at the end of the day.*

As my tram arrives, I hop on, get a window seat and watch the boipara *as it passes by. I wonder, how are Sukumar's and Dev da's experiences different from mine? We all grew up within the* boipara, *making it our own, and now we have different relationships with the space, the books, the unfoldings, the politics and every other aspect that comprises the* boipara. *As I keep thinking about these multiple trajectories of experiences, woven in through stories, moments, pauses and memories, a sense of excitement and anticipation envelops my imagination. The next time I step into the* boipara, *I hope I am able to open myself to newer trajectories, which simultaneously open themselves to me.*

Note

1 There is a complex process of translation as well as interpretation at work here. For me, it was not just his words but the emotions of endearment and a sad note of complaint indicating that I may have forgotten the space where I had spent so much time as a young adult. All these feelings were evident in how he communicated.

References

De Certeau, Michel, & Steven F. Rendall. (2004). From the practice of everyday life (1984). *The City Cultures Reader*, 3, 266.
Deleuze, G. (1995). *Negotiations*. Columbia University Press.
Deleuze, G., & Guattari, F. (1987). *A thousand plateaus: Capitalism and schizophrenia*. University of Minnesota Press.
Gregson, N., & Crewe, L. (2003). *Second-hand cultures*. Berg.

Index